Student Solutions Manual

for

Modern Physics

Third Edition

Raymond A. Serway
Professor Emeritus, James Madison University

Clement J. Moses
Professor Emeritus, Utica College of Syracuse University

Curt A. Moyer
University of North Carolina–Wilmington

THOMSON

BROOKS/COLE

Australia • Canada • Mexico • Singapore • Spain • United Kingdom • United States

Printer: Thomson/West
ISBN-13: 978-0-534-49341-7
ISBN-10: 0-534-49341-6

For more information about our products, contact us at:
Thomson Learning Academic Resource Center
1-800-423-0563

For permission to use material from this text or product, submit a request online at
http://www.thomsonrights.com.
Any additional questions about permissions can be submitted by email to **thomsonrights@thomson.com.**

Thomson Brooks/Cole
10 Davis Drive
Belmont, CA 94002-3098
USA

Asia
Thomson Learning
5 Shenton Way #01-01
UIC Building
Singapore 068808

Australia/New Zealand
Thomson Learning
102 Dodds Street
Southbank, Victoria 3006
Australia

Canada
Nelson
1120 Birchmount Road
Toronto, Ontario M1K 5G4
Canada

Europe/Middle East/South Africa
Thomson Learning
High Holborn House
50/51 Bedford Row
London WC1R 4LR
United Kingdom

Latin America
Thomson Learning
Seneca, 53
Colonia Polanco
11560 Mexico D.F.
Mexico

Spain/Portugal
Paraninfo
Calle/Magallanes, 25
28015 Madrid, Spain

Contents

Chapter 1: Relativity I — 1

Chapter 2: Relativity II — 7

Chapter 3: The Quantum Theory of Light — 13

Chapter 4: The Particle Nature of Matter — 23

Chapter 5: Matter Waves — 33

Chapter 6: Quantum Mechanics in One Dimension — 41

Chapter 7: Tunneling Phenomena — 49

Chapter 8: Quantum Mechanics in Three Dimensions — 53

Chapter 9: Atomic Structure — 59

Chapter 10: Statistical Physics — 65

Chapter 11: Molecular Structure — 75

Chapter 12: The Solid State — 83

Chapter 13: Nuclear Structure — 93

Chapter 14: Nuclear Physics Applications — 105

Chapter 15: Particle Physics — 115

1

Relativity I

1-1 $F = \dfrac{dP}{dt}$. Consider the special case of constant mass. Then, this equation reduces to $\mathbf{F}_A = m\mathbf{a}_A$ in the stationary reference system, and $\mathbf{v}_B = \mathbf{v}_A + \mathbf{v}_{BA}$ where the subscript A indicates that the measurement is made in the laboratory frame, B the moving frame, and $\mathbf{v}_{BA}$ is the velocity of B with respect to A. It is given that $\mathbf{a}_1 = \dfrac{d\mathbf{v}_{BA}}{dt}$. Therefore from differentiating the velocity equation, we have $\mathbf{a}_B = \mathbf{a}_A + \mathbf{a}_1$. Assuming mass is invariant, and the forces are invariant as well, the Newton's law in frame B should be $\sum \mathbf{F} = m\mathbf{a}_A = m\mathbf{a}_B - m\mathbf{a}_1$, which is not simply $m\mathbf{a}_B$. So Newton's second law $\sum \mathbf{F} = m\mathbf{a}_B$ is invalid in frame B. However, we can rewrite it as $\sum \mathbf{F} + m\mathbf{a}_1 = m\mathbf{a}_B$, which compares to $\sum \mathbf{F} + m\mathbf{g} = m\mathbf{a}_B$. It is as if there were a universal gravitational field $\mathbf{g}$ acting on everything. This is the basic idea of the equivalence principle (General Relativity) where an accelerated reference frame is equivalent to a reference frame with a universal gravitation field.

1-3   **IN THE REST FRAME:**
In an elastic collision energy and momentum are conserved.

$$p_i = m_1 v_{1i} + m_2 v_{2i} = (0.3 \text{ kg})(5 \text{ m/s}) + (0.2 \text{ kg})(-3 \text{ m/s}) = 0.9 \text{ kg} \cdot \text{m/s}$$
$$p_f = m_1 v_{1f} + m_2 v_{2f}$$

This equation has two unknowns, therefore, apply the conservation of kinetic energy $E_i = E_f = \dfrac{1}{2} m_1 v_{1i}^2 + \dfrac{1}{2} m_2 v_{2i}^2 = \dfrac{1}{2} m_1 v_{1f}^2 + \dfrac{1}{2} m_2 v_{2f}^2$ and conservation of momentum one finds that $v_{1f} = -1.31$ m/s and $v_{2f} = 6.47$ m/s or $v_{1f} = -1.56$ m/s and $v_{2f} = 6.38$ m/s. The difference in values is due to the rounding off errors in the numerical calculations of the mathematical quantities. If these two values are averaged the values are $v_{1f} = -1.4$ m/s and $v_{2f} = 6.6$ m/s, $p_f = 0.9$ kg·m/s. Thus, $p_i = p_f$.

IN THE MOVING FRAME:
Make use of the Galilean velocity transformation equations. $p_i' = m_1 v_{1i}' + m_2 v_{2i}'$; where $v_{1i}' = v_{1i} - v' = 5$ m/s $- (-2$ m/s$) = 7$ m/s. Similarly, $v_{2i}' = -1$ m/s and $p_i' = 1.9$ kg·m/s. To find p_f' use $v_{1f}' = v_{1i} - v'$ and $v_{2f}' = v_{2i} - v'$ because the prime system is now moving to the left. Using these results give $p_f' = 1.9$ kg·m/s.

1-5   This is a case of dilation. $T = \gamma T'$ in this problem with the proper time $T' = T_0$

$$T = \left[1 - \left(\frac{v}{c}\right)^2\right]^{-1/2} T_0 \Rightarrow \frac{v}{c} = \left[1 - \left(\frac{T_0}{T}\right)^2\right]^{1/2};$$

in this case $T = 2T_0$, $v = \left\{1 - \left[\frac{L_0/2}{L_0}\right]^2\right\}^{1/2} = \left[1 - \left(\frac{1}{4}\right)\right]^{1/2}$ therefore $v = 0.866c$.

1-7   The problem is solved by using time dilation. This is also a case of $v \ll c$ so the binomial expansion is used $\Delta t = \gamma \Delta t' \cong \left[1 + \frac{v^2}{2c^2}\right]\Delta t'$, $\Delta t - \Delta t' = \frac{v^2 \Delta t'}{2c^2}$; $v = \left[\frac{2c^2(\Delta t - \Delta t')}{\Delta t'}\right]^{1/2}$;

$\Delta t = (24 \text{ h/day})(3\,600 \text{ s/h}) = 86\,400 \text{ s}$; $\Delta t = \Delta t' - 1 = 86\,399 \text{ s}$;

$$v = \left[\frac{2(86\,400 \text{ s} - 86\,399 \text{ s})}{86\,399 \text{ s}}\right]^{1/2} = 0.004\,8c = 1.44 \times 10^6 \text{ m/s}.$$

1-9 $L_{earth} = \dfrac{L'}{\gamma}$

$L_{earth} = L'\left[1 - \frac{v^2}{c^2}\right]^{1/2}$, L', the proper length so $L_{earth} = L = L\left[1 - (0.9)^2\right]^{1/2} = 0.436L$.

1-11 $\Delta t = \gamma \Delta t'$

$$\Delta t = \Delta t'\left(1 - \frac{v^2}{c^2}\right)^{-1/2} \cong \left(1 + \frac{v^2}{2c^2}\right)\Delta t' \cong \left[1 + \frac{\left(4.0 \times 10^2 \text{ m/s}\right)^2}{2\left(3.0 \times 10^8 \text{ m/s}\right)^2}\right](3\,600 \text{ s})$$

$\cong \left(1 + 8.89 \times 10^{-13}\right)(3\,600 \text{ s}) = \left(3\,600 + 3.2 \times 10^{-9}\right) \text{ s}$

$\Delta t - \Delta t' \cong 3.2 \text{ ns}$. (Moving clocks run slower.)

1-13   (a) $\tau = \gamma \tau' = \left[1 - (0.95)^2\right]^{-1/2}(2.2 \text{ } \mu s) = 7.05 \text{ } \mu s$

        (b) $\Delta t' = \dfrac{d}{0.95c} = \dfrac{3 \times 10^3 \text{ m}}{0.95c} = 1.05 \times 10^{-5} \text{ s}$, therefore,

$$N = N_0 \exp\left(-\frac{\Delta t}{\tau}\right) = \left(5 \times 10^4 \text{ muons}\right)\exp(-1.487) \approx 1.128 \times 10^4 \text{ muons}.$$

1-15   (a)   For a receding source we replace v by $-v$ in Equation 1.15 and obtain:

$$f_{ob} = \left\{\frac{[c - v]^{1/2}}{[c + v]^{1/2}}\right\}f_{source} = \left\{\frac{[1 - v/c]^{1/2}}{[1 + v/c]^{1/2}}\right\}f_{source} \cong \left(1 - \frac{v}{2c}\right)\left(1 - \frac{v}{2c}\right)f_{source}$$

$$\cong \left(1 - \frac{v}{c} + \frac{v^2}{4c^2}\right)f_{source} \cong \left(1 - \frac{v}{c}\right)f_{source}$$

where we have used the binomial expansion and have neglected terms of second and higher order in $\dfrac{v}{c}$. Thus, $\dfrac{\Delta f}{f_{source}} = \dfrac{f_{ob} - f_{source}}{f_{source}} = -\dfrac{v}{c}$.

(b)   From the relations $f = \dfrac{c}{\lambda}$, $\dfrac{df}{d\lambda} = -\dfrac{c}{\lambda^2}$ we find $\dfrac{df}{f} = -\dfrac{c/\lambda^2}{c/\lambda}d\lambda$, or $\dfrac{\Delta\lambda}{\lambda} = -\dfrac{\Delta f}{f} = \dfrac{v}{c}$.

(c)   Assuming $v \ll c$, $\dfrac{v}{c} \cong \dfrac{\Delta\lambda}{\lambda}$, or $v \cong \left(\dfrac{\Delta\lambda}{\lambda}\right)c = \left(\dfrac{20\text{ nm}}{397\text{ nm}}\right)c = 0.050c = 1.5\times10^7$ m/s.

1-17   (a)   Galaxy A is approaching and as a consequence it exhibits blue shifted radiation. From Example 1.6, $\beta = \dfrac{v}{c} = \dfrac{\lambda_{source}^2 - \lambda_{obs}^2}{\lambda_{source}^2 + \lambda_{obs}^2}$ so that $\beta = \dfrac{(550\text{ nm})^2 - (450\text{ nm})^2}{(550\text{ nm})^2 + (450\text{ nm})^2} = 0.198$. Galaxy A is approaching at $v = 0.198c$.

(b)   For a red shift, B is receding. $\beta = \dfrac{v}{c} = \dfrac{\lambda_{source}^2 - \lambda_{obs}^2}{\lambda_{source}^2 + \lambda_{obs}^2}$ so that

$\beta = \dfrac{(700\text{ nm})^2 - (550\text{ nm})^2}{(700\text{ nm})^2 + (550\text{ nm})^2} = 0.237$. Galaxy B is receding at $v = 0.237c$.

1-19 $u_{XA} = -u_{XB}$; $u'_{XA} = 0.7c = \dfrac{u_{XA} - u_{XB}}{1 - u_{XA}u_{XB}/c^2}$; $0.70c = \dfrac{2u_{XA}}{1 + (u_{XA}/c)^2}$ or $0.70u_{XA}^2 - 2cu_{XA} + 0.7c^2 = 0$.
Solving this quadratic equation one finds $u_{XA} = 0.41c$ therefore $u_{XB} = -u_{XA} = -0.41c$.

1-21 $u'_X = \dfrac{u_X - v}{1 - u_X v/c^2} = \dfrac{0.50c - 0.80c}{1 - (0.50c)(0.80c)/c^2} = -0.50c$

1-23   (a)   Let event 1 have coordinates $x_1 = y_1 = z_1 = t_1 = 0$ and event 2 have coordinates $x_2 = 100$ mm, $y_2 = z_2 = t_2 = 0$. In S', $x'_1 = \gamma(x_1 - vt_1) = 0$, $y'_1 = y_1 = 0$, $z'_1 = z_1 = 0$, and $t'_1 = \gamma\left[t_1 - \left(\dfrac{v}{c^2}\right)x_1\right] = 0$, with $\gamma = \left[1 - \dfrac{v^2}{c^2}\right]^{-1/2}$ and so $\gamma = \left[1 - (0.70)^2\right]^{-1/2} = 1.40$. In system S', $x'_2 = \gamma(x_2 - vt_2) = 140$ m, $y'_2 = z'_2 = 0$, and

$$t'_2 = \gamma\left[t_2 - \left(\dfrac{v}{c^2}\right)x_2\right] = \dfrac{(1.4)(-0.70)(100\text{ m})}{3.00\times10^8\text{ m/s}} = -0.33\ \mu s.$$

(b) $\Delta x' = x'_2 - x'_1 = 140$ m

(c)   Events are not simultaneous in S', event 2 occurs 0.33 μs earlier than event 1.

1-25   We find Carpenter's speed: $\dfrac{mGM}{r^2} = \dfrac{mv^2}{r}$

$$v = \left[\dfrac{GM}{R+h}\right]^{1/2} = \left[\dfrac{(6.67\times10^{-11})(5.98\times10^{24})}{6.37\times10^6 + 0.16\times10^6}\right]^{1/2} = 7.82\text{ km/s}.$$

Then the period of one orbit is $T = \dfrac{2\pi(R+h)}{v} = \dfrac{2\pi(6.53\times10^6)}{7.82\times10^3} = 5.25\times10^3$ s.

(a)   The time difference for 22 orbits is $\Delta t - \Delta t' = (\gamma - 1)\Delta t' = \left[\left(1 - \dfrac{v^2}{c^2}\right)^{-1/2} - 1\right](22)(T)$.

Using the binomial expansion one obtains

$$\left(1 + \frac{1}{2}\frac{v^2}{c^2} - 1\right)(22)(T) = \frac{1}{2}\left[\frac{7.82 \times 10^3 \text{ m/s}}{3 \times 10^8 \text{ m/s}}\right](22)(5.5 \times 10^3 \text{ s}) = 39.2 \ \mu s.$$

(b)   For one orbit, $\Delta t - \Delta t' = \dfrac{39.2 \ \mu s}{22} = 1.78 \ \mu s \approx 2 \ \mu s$. The press report is accurate to one significant figure.

1-27   For the pion to travel 10 m in time Δt in our frame,

$$10 \text{ m} = v\Delta t = v(\gamma\,\Delta t') = v\left(26 \times 10^{-9} \text{ s}\right)\left[1 - \left(\frac{v}{c}\right)^2\right]^{-1/2}$$

$$\left(3.85 \times 10^8 \text{ m/s}\right)^2\left(1 - \frac{v^2}{c^2}\right) = v^2$$

$$1.46 \times 10^{17} \text{ m}^2/\text{s}^2 = v^2(1 + 1.64)$$

$$v = 2.37 \times 10^8 \text{ m/s} = 0.789c$$

1-29   (a)   A spaceship, reference frame S', moves at speed v relative to the Earth, whose reference frame is S. The space ship then launches a shuttle craft with velocity v in the forward direction. The pilot of the shuttle craft then fires a probe with velocity v in the forward direction. Use the relativistic compounding of velocities as well as its inverse transformation: $u'_x = \dfrac{u_x - v}{1 - \left(u_x v/c^2\right)}$, and its inverse $u_x = \dfrac{u'_x + v}{1 + \left(u'_x v/c^2\right)}$. The above variables are defined as: v is the spaceship's velocity relative to S, u'_x is the velocity of the shuttle craft relative to S', and u_x is the velocity of the shuttle craft relative to S. Setting u'_x equal to v, we find the velocity of the shuttle craft relative to the Earth to be: $u_x = \dfrac{2v}{1 + (v/c)^2}$.

(b)   If we now take S to be the shuttle craft's frame of reference and S' to be that of the probe whose speed is v relative to the shuttle craft, then the speed of the probe relative to the spacecraft will be, $u'_x = \dfrac{2v}{1 + (v/c)^2}$. Adding the speed relative to S yields:

$u_x = \left[\dfrac{3 + (v/c)^2}{1 + 2(v/c)^2}\right] = \dfrac{3v + v^3/c^3}{1 + 2v^2/c^2}$. Using the Galilean transformation of velocities, we see that the spaceship's velocity relative to the Earth is v, the velocity of the shuttle craft relative to the space ship is v and therefore the velocity of the shuttle craft relative to the Earth must be $2v$ and finally the speed of the probe must be $3v$. In the limit of low $\left(\dfrac{v}{c}\right)^2$, u_x reduces to $3v$. On the other hand, using relativistic addition of velocities, we find that $u_x = c$ when $v \rightarrow c$.

1-31   In this case, the proper time is T_0 (the time measured by the students using a clock at rest relative to them). The dilated time measured by the professor is: $\Delta t = \gamma T_0$ where $\Delta t = T + t$. Here T is the time she waits before sending a signal and t is the time required for the signal to reach the students. Thus we have: $T + t = \gamma T_0$. To determine travel time t, realize that the distance the students will have moved beyond the professor before the signal reaches them is: $d = v(T + t)$. The time required for the signal to travel this distance is: $t = \dfrac{d}{c} = \dfrac{v}{c}(T + t)$. Solving

for t gives: $t = \left(\dfrac{v}{c}\right)T\left(1 - \dfrac{v}{c}\right)^{-1}$. Substituting this into the above equation for $(T + t)$ yields:

$T + \left(\dfrac{v}{c}\right)T\left(1 - \dfrac{v}{c}\right)^{-1} = \gamma T_0$, or $T\left(1 - \dfrac{v}{c}\right)^{-1} = \gamma T_0$. Using the expression for γ this becomes:

$T = \left(1 - \dfrac{v}{c}\right)\left[1 - \left(\dfrac{v}{c}\right)^2\right]^{-1/2} T_0$, or $T = T_0\left(1 - \dfrac{v}{c}\right)\left[1 - \left(\dfrac{v}{c}\right)^2\right]^{-1/2} = T_0\left[\left(1 - \dfrac{v}{c}\right)\left(1 + \dfrac{v}{c}\right)^{-1}\right]^{1/2}$.

1-33   (a)   We in the spaceship moving past the hermit do not calculate the explosions to be simultaneous. We measure the distance we have traveled from the Sun as

$$L = L_p\sqrt{1 - \left(\frac{v}{c}\right)^2} = (6.00 \text{ ly})\sqrt{1 - (0.800)^2} = 3.60 \text{ ly}.$$

We see the Sun flying away from us at $0.800c$ while the light from the Sun approaches at $1.00c$. Thus, the gap between the Sun and its blast wave has opened at $1.80c$, and the time we calculate to have elapsed since the Sun exploded is $\dfrac{3.60 \text{ ly}}{1.80c} = 2.00 \text{ yr}$. We see Tau Ceti as moving toward us at $0.800c$, while its light approaches at $1.00c$, only $0.200c$ faster. We measure the gap between that star and its blast wave as 3.60 ly and growing at $0.200c$. We calculate that it must have been opening for $\dfrac{3.60 \text{ ly}}{0.200c} = 18.0 \text{ yr}$ and conclude that $\boxed{\text{Tau Ceti exploded 16.0 years before the Sun}}$.

(b)   Consider a hermit who lives on an asteroid halfway between the Sun and Tau Ceti, stationary with respect to both. Just as our spaceship is passing him, he also sees the blast waves from both explosions. Judging both stars to be stationary, this observer concludes that $\boxed{\text{the two stars blew up simultaneously}}$.

1-35   In the Earth frame, Speedo's trip lasts for a time $\Delta t = \dfrac{\Delta x}{v} = \dfrac{20.0 \text{ ly}}{0.950 \text{ ly/yr}} = 21.05$ Speedo's age advances only by the proper time interval: $\Delta t_p = \dfrac{\Delta t}{\gamma} = 21.05 \text{ yr}\sqrt{1 - 0.95^2} = 6.574 \text{ yr}$ during his trip. Similarly for Goslo, $\Delta t_p = \dfrac{\Delta x}{v}\sqrt{1 - \dfrac{v^2}{c^2}} = \dfrac{20.0 \text{ ly}}{0.750 \text{ ly/yr}}\sqrt{1 - 0.75^2} = 17.64 \text{ yr}$. While Speedo has landed on Planet X and is waiting for his brother, he ages by

$$\frac{20.0 \text{ ly}}{0.750 \text{ ly/yr}} - \frac{0.20 \text{ ly}}{0.950 \text{ ly/yr}}\sqrt{1 - 0.75^2} = 17.64 \text{ yr}.$$

Then Goslo ends up older by $17.64 \text{ yr} - (6.574 \text{ yr} + 5.614 \text{ yr}) = 5.45 \text{ yr}$.

1-37   Einstein's reasoning about lightning striking the ends of a train shows that the moving observer sees the event toward which she is moving, event B, as occurring first. We may take the S-frame coordinates of the events as $(x = 0, y = 0, z = 0, t = 0)$ and $(x = 100$ m, $y = 0, z = 0, t = 0)$. Then the coordinates in S' are given by Equations 1.23 to 1.27. Event A is at $(x' = 0, y' = 0, z' = 0, t' = 0)$. The time of event B is:

$$t' = \gamma\left(t - \frac{v}{c^2}x\right) = \frac{1}{\sqrt{1 - 0.8^2}}\left(0 - \frac{0.8c}{c^2}(100 \text{ m})\right) = 1.667\left(\frac{80 \text{ m}}{3 \times 10^8 \text{ m/s}}\right) = -4.44 \times 10^{-7} \text{ s}.$$

The time elapsing before A occurs is 444 ns.

1-39   (a)   For the satellite $\sum F = ma$: $\dfrac{GM_E m}{r^2} = \dfrac{mv^2}{r} = \dfrac{m}{r}\left(\dfrac{2\pi r}{T}\right)^2$

$$GM_E T^2 = 4\pi^2 r^3$$

$$r = \left(\frac{6.67 \times 10^{-11} \text{ N} \cdot \text{m}^2 (5.98 \times 10^{24} \text{ kg})(43\,080 \text{ s})^2}{\text{kg}^2 \, 4\pi^2}\right)^{1/3} = 2.66 \times 10^7 \text{ m}$$

(b) $v = \dfrac{2\pi r}{T} = \dfrac{2\pi(2.66 \times 10^7 \text{ m})}{43\,080 \text{ s}} = 3.87 \times 10^3 \text{ m/s}$

(c)   The small fractional decrease in frequency received is equal in magnitude to the fractional increase in period of the moving oscillator due to time dilation:

$$\text{fractional change in } f = -(\gamma - 1) = -\left[\frac{1}{\sqrt{1 - (3.87 \times 10^3 / 3 \times 10^8)^2}} - 1\right]$$

$$= 1 - \left(1 - \frac{1}{2}\left[-\left(\frac{3.87 \times 10^3}{3 \times 10^8}\right)^2\right]\right) = -8.34 \times 10^{-11}$$

(d)   The orbit altitude is large compared to the radius of the Earth, so we must use $U_g = -\dfrac{GM_E m}{r}$.

$$\Delta U_g = -\frac{6.67 \times 10^{-11} \text{ N} \cdot \text{m}^2 (5.98 \times 10^{24} \text{ kg})m}{\text{kg}^2 \, 2.66 \times 10^7 \text{ m}} + \frac{6.67 \times 10^{-11} \text{ N} \cdot \text{m}^2 (5.98 \times 10^{24} \text{ kg})m}{\text{kg}^2 \, 6.37 \times 10^6 \text{ m}}$$

$$= 4.76 \times 10^7 \text{ J/kg} \, m$$

$$\frac{\Delta f}{f} = \frac{\Delta U_g}{mc^2} = \frac{4.76 \times 10^7 \text{ m}^2/\text{s}^2}{(3 \times 10^8 \text{ m/s})^2} = +5.29 \times 10^{-10}$$

(e) $-8.34 \times 10^{-11} + 5.29 \times 10^{-10} = +4.46 \times 10^{-10}$

2

Relativity II

2-1 $p = \dfrac{mv}{\left[1-(v^2/c^2)\right]^{1/2}}$

(a) $p = \dfrac{(1.67 \times 10^{-27} \text{ kg})(0.01c)}{\left[1-(0.01c/c)^2\right]^{1/2}} = 5.01 \times 10^{-21} \text{ kg}\cdot\text{m/s}$

(b) $p = \dfrac{(1.67 \times 10^{-27} \text{ kg})(0.5c)}{\left[1-(0.5c/c)^2\right]^{1/2}} = 2.89 \times 10^{-19} \text{ kg}\cdot\text{m/s}$

(c) $p = \dfrac{(1.67 \times 10^{-27} \text{ kg})(0.9c)}{\left[1-(0.9c/c)^2\right]^{1/2}} = 1.03 \times 10^{-18} \text{ kg}\cdot\text{m/s}$

(d) $\dfrac{1.00 \text{ MeV}}{c} = \dfrac{1.602 \times 10^{-13} \text{ J}}{2.998 \times 10^8 \text{ m/s}} = 5.34 \times 10^{-22} \text{ kg}\cdot\text{m/s}$ so for (a)

$$p = \dfrac{(5.01 \times 10^{-21} \text{ kg}\cdot\text{m/s})(100 \text{ MeV}/c)}{5.34 \times 10^{-22} \text{ kg}\cdot\text{m/s}} = 9.38 \text{ MeV}/c$$

Similarly, for (b) $p = 540 \text{ MeV}/c$ and for (c) $p = 1\,930 \text{ MeV}/c$.

2-3 As **F** is parallel to **v**, scalar equations are used. Relativistic momentum is given by
$p = \gamma mv = \dfrac{mv}{\left[1-(v/c)^2\right]^{1/2}}$, and relativistic force is given by

$$F = \frac{dp}{dt} = \frac{d}{dt}\left\{\frac{mv}{\left[1-(v/c)^2\right]^{1/2}}\right\}$$

$$F = \frac{dp}{dt} = \frac{m}{\left[1-(v^2/c^2)\right]^{3/2}}\left(\frac{dv}{dt}\right)$$

2-5   This is the case where we use the relativistic form of Newton's second law, but unlike Problem 2-3 in which **F** is parallel to **v**, here **F** is perpendicular to **v** and $\mathbf{F} = \dfrac{d\mathbf{p}}{dt}$ so that

$$\mathbf{F} = q\mathbf{v} \times \mathbf{B} = \frac{d\mathbf{p}}{dt} = \frac{d}{dt}\left\{\frac{mv}{\sqrt{1-(v/c)^2}}\right\}.$$

Assuming that **B** is perpendicular to the plane of the orbit of q, the force is radially inward, and we find

$$\mathbf{F} = q\mathbf{v}\mathbf{B}\big|_{\text{radial}} = \frac{d}{dt}\left\{\frac{mv}{\sqrt{1-(v/c)^2}}\right\}.$$

As the force is perpendicular to **v**, it does no work on the charge and the magnitude (but not the direction) of **v** remains constant in time. Thus,

$$\frac{d}{dt}\left\{\frac{mv}{\sqrt{1-(v/c)^2}}\right\} = \frac{m}{\sqrt{1-(v/c)^2}}\frac{dv}{dt}.$$

Identifying $\left(\dfrac{dv}{dt}\right)$ as the centripetal acceleration where the scalar equation $\dfrac{dv}{dt} = \left(\dfrac{v^2}{r}\right)_{\text{radial}}$

gives $qvB\big|_{\text{radial}} = \left[\dfrac{m}{1-v^2/c^2}\right]^{1/2}\left(\dfrac{v^2}{r}\right)\bigg|_{\text{radial}}$ or $v = \left(\dfrac{qBr}{m}\right)\left(1-\dfrac{v^2}{c^2}\right)^{1/2}$. Finally, the period T is $\dfrac{2\pi r}{v}$

and $T = \dfrac{2\pi r}{(qBr/m)(1-v^2/c^2)^{1/2}} = \dfrac{2\pi m}{(qB)(1-v^2/c^2)^{1/2}}$. As $f = \dfrac{1}{T}$, $f = \left(\dfrac{qB}{2\pi m}\right)\left(1-\dfrac{v^2}{c^2}\right)^{1/2}$.

2-7 $E = \gamma mc^2$, $p = \gamma mu$; $E^2 = (\gamma mc^2)^2$; $p^2 = (\gamma mu)^2$;

$$E^2 - p^2c^2 = (\gamma mc^2)^2 - (\gamma mu)^2 c^2 = \gamma^2\left\{(mc^2)^2 - (mc)^2 u^2\right\}$$

$$= (mc^2)^2\left(1-\frac{u^2}{c^2}\right)\left(1-\frac{u^2}{c^2}\right)^{-1} = (mc^2)^2 \text{ Q.E.D.}$$

$$E^2 = p^2c^2 + (mc^2)^2$$

2-9   (a)   When $K = (\gamma - 1)mc^2 = 5mc^2$, $\gamma = 6$ and $E = \gamma mc^2 = 6(0.5110 \text{ MeV}) = 3.07 \text{ MeV}$.

(b) $\dfrac{1}{\gamma} = \left[1 - \left(\dfrac{v}{c}\right)^2\right]^{1/2}$ and $v = c\left[1 - \left(\dfrac{1}{\gamma}\right)^2\right]^{1/2} = c\left[1 - \left(\dfrac{1}{6}\right)^2\right]^{1/2} = 0.986c$

2-11   (a) $K = 50 \times 10^9 \text{ eV}; \; mc^2 = 938.27 \text{ MeV};$

$$E = K + mc^2 = \left(50 \times 10^9 \text{ eV}\right) + \left(938.27 \times 10^6 \text{ eV}\right) = 50\,938.3 \text{ MeV}$$

$$E^2 = p^2 c^2 + m^2 c^4 \Rightarrow p = \left[\frac{E^2 - m^2 c^4}{c^2}\right]^{1/2}$$

$$p = \frac{\left[(50\,938.3 \text{ MeV})^2 - (938.27 \text{ MeV})^2\right]^{1/2}}{c} = 5.09 \times 10^{10} \text{ eV}/c$$

$$= \frac{5.09 \times 10^{10} \text{ eV}}{3 \times 10^8 \text{ m/s}} \left(1.6 \times 10^{-19} \text{ J/eV}\right) = 2.71 \times 10^{-17} \text{ kg} \cdot \text{m/s}$$

(b) $$E = \gamma mc^2 = \frac{mc^2}{\left[1 - (v/c)^2\right]^{1/2}} \Rightarrow v = c\left[1 - \left(\frac{mc^2}{E}\right)^2\right]^{1/2}$$

$$= \left(3 \times 10^8 \text{ m/s}\right)\left[1 - \left(\frac{938.27 \text{ MeV}}{50\,938.3 \text{ MeV}}\right)^2\right]^{1/2} = 2.999\,5 \times 10^8 \text{ m/s}$$

2-13   (a) $E = 400 mc^2 = \gamma mc^2$

$\gamma = \left(1 - \dfrac{v^2}{c^2}\right)^{-1/2} = 400$

$\left(1 - \dfrac{v^2}{c^2}\right) = \left(\dfrac{1}{400}\right)^2$

$\dfrac{v}{c} = \left[1 - \dfrac{1}{400^2}\right]^{1/2}$

$v = 0.999\,997c$

(b) $K = E - mc^2 = (400 - 1)mc^2 = 399 mc^2 = (399)(938.3 \text{ MeV}) = 3.744 \times 10^5 \text{ MeV}$

2-15   (a) $K = \gamma mc^2 - mc^2 = Vq$ and so, $\gamma^2 = \left(1 + \dfrac{Vq}{mc^2}\right)^2$ and $\dfrac{v}{c} = \left\{1 - \left(1 + \dfrac{Vq}{mc^2}\right)^{-2}\right\}^{1/2}$

$$\frac{v}{c} = \left\{1 - \frac{1}{1 + \left(5.0 \times 10^4 \text{ eV}/0.511 \text{ MeV}\right)^2}\right\}^{1/2} = 0.412\,7$$

or $v = 0.413c$.

(b) $K = \dfrac{1}{2}mv^2 = Vq$

$v = \left(\dfrac{2Vq}{m}\right)^{1/2} = \left\{\dfrac{2\left(5.0 \times 10^4 \text{ eV}\right)}{0.511 \text{ MeV}/c^2}\right\}^{1/2} = 0.442c$

(c)   The error in using the classical expression is approximately $\dfrac{3}{40} \times 100\%$ or about 7.5% in speed.

2-17 $\Delta m = m_{\text{Ra}} - m_{\text{Rn}} - m_{\text{He}}$ (an atomic unit of mass, the u, is one-twelfth the mass of the ^{12}C atom or $1.660\,54 \times 10^{-27}$ kg)

$$\Delta m = (226.025\,4 - 22.017\,5 - 4.002\,6)\,\text{u} = 0.005\,3\,\text{u}$$
$$E = (\Delta m)(931\ \text{MeV/u}) = (0.005\,3\ \text{u})(931\ \text{MeV/u}) = 4.9\ \text{MeV}$$

2-19 $\Delta m = 6m_p + 6m_n - m_C = [6(1.007\,276) + 6(1.008\,665) - 12]\,\text{u} = 0.095\,646\,\text{u}$,

$$\Delta E = (931.49\ \text{MeV/u})(0.095\,646\ \text{u}) = 89.09\ \text{MeV}.$$

Therefore the energy per nucleon $= \dfrac{89.09\ \text{MeV}}{12} = 7.42\ \text{MeV}$.

2-21

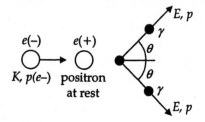

Conservation of mass-energy requires $K + 2mc^2 = 2E$ where K is the electron's kinetic energy, m is the electron's mass, and E is the gamma ray's energy.

$$E = \frac{K}{2} + mc^2 = (0.500 + 0.511)\ \text{MeV} = 1.011\ \text{MeV}.$$

Conservation of momentum requires that $p_{e^-} = 2p\cos\theta$ where p_{e^-} is the initial momentum of the electron and p is the gamma ray's momentum, $\dfrac{E}{c} = 1.011\ \text{MeV}/c$. Using $E_{e^-}^2 = p_{e^-}^2 c^2 + (mc^2)^2$ where E_{e^-} is the electron's total energy, $E_{e^-} = K + mc^2$, yields

$$p_{e^-} = \frac{1}{c}\sqrt{K^2 + 2Kmc^2} = \frac{\sqrt{(1.00)^2 + 2(1.00)(0.511)}\ \text{MeV}}{c} = 1.422\ \text{MeV}/c.$$

Finally, $\cos\theta = \dfrac{p_{e^-}}{2p} = 0.703$; $\theta = 45.3°$.

2-23   In this problem, M is the mass of the initial particle, m_l is the mass of the lighter fragment, v_l is the speed of the lighter fragment, m_h is the mass of the heavier fragment, and v_h is the speed of the heavier fragment. Conservation of mass-energy leads to

$$Mc^2 = \frac{m_l c^2}{\sqrt{1 - v_l^2/c^2}} + \frac{m_h c^2}{\sqrt{1 - v_h^2/c^2}}$$

From the conservation of momentum one obtains

$$(m_l)(0.987c)(6.22) = (m_h)(0.868c)(2.01)$$

$$m_l = \frac{(m_h)(0.868c)(2.01)}{(0.987)(6.22)} = 0.284 m_h$$

Substituting in this value and numerical quantities in the mass-energy conservation equation, one obtains 3.34×10^{-27} kg $= 6.22 m_l + 2.01 m_h$ which in turn gives

3.34×10^{-27} kg $= (6.22)(0.284) m_l + 2.01 m_h$ or $m_h = \dfrac{3.34 \times 10^{-27} \text{ kg}}{3.78} = 8.84 \times 10^{-28}$ kg and

$m_l = (0.284) m_h = 2.51 \times 10^{-28}$ kg.

2-25   (a)   The x component of the gravitational force between a light particle of mass m and the Sun is given by $F_x = \dfrac{GM_S m}{r^2} \sin\phi = \dfrac{GM_S mb}{\left(b^2 + y^2\right)^{3/2}}$. The change in momentum in the x direction is given by $\Delta p_x = \int_{-\infty}^{\infty} F_x dt = \int_{-\infty}^{\infty} \dfrac{GM_S mb}{\left(b^2 + y^2\right)^{3/2}} dt$. To convert dt to dy, assume the deflection is very small and that the position of the light particle is given by $y = -ct$ for $x = 0$. Thus $dt = -\dfrac{dy}{c}$ and we get

$$\Delta p_x = -\frac{GM_S mb}{c} \int_{+\infty}^{-\infty} \frac{dy}{\left(b^2 + y^2\right)^{3/2}} = \frac{2GM_S mb}{c} \int_{0}^{+\infty} \frac{dy}{\left(b^2 + y^2\right)^{3/2}} = \frac{2GM_S mb}{c} \left. \frac{y}{b^2\left(y^2 + b^2\right)^{1/2}} \right|_0^{\infty}$$

$$= \frac{2GM_S mb}{c}\left(\frac{1}{b^2}\right) = \frac{2GM_S m}{cb}$$

From Figure P2.25(b), $\theta \cong \dfrac{\Delta p_x}{mc}$ so we find $\theta \cong \dfrac{2GM_S m}{cb(mc)} = \dfrac{2GM_S}{bc^2}$.

(b)   For $b = R_S = 6.96 \times 10^8$ m and $M_S = 1.99 \times 10^{30}$ kg

$$\theta = \frac{2\left(6.67 \times 10^{-11} \text{ N} \cdot \text{m}^2/\text{kg}^2\right)\left(1.99 \times 10^{30} \text{ kg}\right)}{\left(6.96 \times 10^8 \text{ m}\right)\left(3.00 \times 10^8 \text{ m/s}\right)^2} = 4.24 \times 10^{-6} \text{ rad} = 2.43 \times 10^{-4} \text{ deg}$$

2-27   If the energy required to remove a mass m from the surface is equal to its rest energy mc^2, then $\dfrac{GM_S m}{R_g} = mc^2$ and $R_g = \dfrac{GM_S}{c^2} = \dfrac{\left(6.67 \times 10^{-11} \text{ N} \cdot \text{m}^2/\text{kg}^2\right)\left(1.99 \times 10^{30} \text{ kg}\right)}{\left(3.00 \times 10^8 \text{ m/s}\right)^2}$,

$R_g = 1.47 \times 10^3$ m $= 1.47$ km.

2-29   The energy of the first fragment is given by $E_1^2 = p_1^2 c^2 + \left(m_1 c^2\right)^2 = (1.75 \text{ MeV})^2 + (1.00 \text{ MeV})^2$; $E_1 = 2.02$ MeV. For the second, $E_2^2 = (2.00 \text{ MeV})^2 + (1.50 \text{ MeV})^2$; $E_2 = 2.50$ MeV.

(a)   Energy is conserved, so the unstable object had $E = 4.52$ MeV. Each component of momentum is conserved, so the original object moved with

$$p^2 = p_x^2 + p_y^2 = \left(\frac{1.75 \text{ MeV}}{c}\right)^2 + \left(\frac{2.00 \text{ MeV}}{c}\right)^2.$$

Then for $(4.52 \text{ MeV})^2 = (1.75 \text{ MeV})^2 + (2.00 \text{ MeV})^2 + \left(mc^2\right)^2$; $m = 3.65 \text{ MeV}/c^2$.

(b)   Now $E = \gamma mc^2$ gives $4.52 \text{ MeV} = \dfrac{1}{\sqrt{1 - v^2/c^2}} 3.65 \text{ MeV}$; $1 - \dfrac{v^2}{c^2} = 0.654$ and $v = 0.589c$.

2-31   Conservation of momentum γmu:

$$\frac{mu}{\sqrt{1 - u^2/c^2}} + \frac{m(-u)}{3\sqrt{1 - u^2/c^2}} = \frac{Mv_f}{\sqrt{1 - v_f^2/c^2}} = \frac{2mu}{3\sqrt{1 - u^2/c^2}}.$$

Conservation of energy γmc^2:

$$\frac{mc^2}{\sqrt{1 - u^2/c^2}} + \frac{mc^2}{3\sqrt{1 - u^2/c^2}} = \frac{Mc^2}{\sqrt{1 - v_f^2/c^2}} = \frac{4mc^2}{3\sqrt{1 - u^2/c^2}}.$$

To start solving we can divide: $v_f = \dfrac{2u}{4} = \dfrac{u}{2}$. Then

$$\frac{M}{\sqrt{1 - u^2/4c^2}} = \frac{4m}{3\sqrt{1 - u^2/c^2}} = \frac{M}{(1/2)\sqrt{4 - u^2/c^2}}$$

$$M = \frac{2m\sqrt{4 - u^2/c^2}}{3\sqrt{1 - u^2/c^2}}$$

Note that when $v \ll c$, this reduces to $M = \dfrac{4m}{3}$, in agreement with the classical result.

2-33   The energy that arrives in one year is

$$E = \mathscr{P}\,\Delta t = \left(1.79 \times 10^{17} \text{ J/s}\right)\left(3.16 \times 10^7 \text{ s}\right) = 5.66 \times 10^{24} \text{ J}.$$

Thus, $m = \dfrac{E}{c^2} = \dfrac{5.66 \times 10^{24} \text{ J}}{\left(3.00 \times 10^8 \text{ m/s}\right)^2} = 6.28 \times 10^7 \text{ kg}.$

3

The Quantum Theory of Light

3-1   (a)

$$E2\pi r = \pi r^2 \left(\frac{dB}{dt}\right)$$

$$E = \left(\frac{r}{2}\right)\left(\frac{dB}{dt}\right)$$

  (b)   If r remains constant, then: $E = Eq = \left(\frac{r}{2}\right)\left(\frac{dB}{dt}\right)e$ so that $Fdt = \left(\frac{r}{2}\right)\left(\frac{dB}{dt}\right)dt = m_e dv$, or

$$dv = \left(\frac{re}{2m_e}\right)dB$$

$$\int_v^{v+\Delta v} dv = \left(\frac{er}{2m_e}\right)\int_0^B dB$$

$$\Delta v = \frac{erB}{2m_e}$$

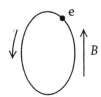

  (c)

$$\Delta\omega = \frac{\Delta v}{r} = \frac{eB}{2m_e} = \left(1.6\times10^{-19}\ C\right)\frac{1\ T}{2}\left(9.1\times10^{-31}\ kg\right) = 8.8\times10^{10}\ rad/sec$$

$$\omega = 2\pi f = \frac{2\pi c}{\lambda} = 2\pi\frac{\left(3.0\times10^8\ m/s\right)}{\left(500\times10^{-9}\ m\right)} = 3.8\times10^{15}\ rad/sec;\quad \frac{\Delta\omega}{\omega} = 2.3\times10^{-5}$$

  (d)   For the ω_0 line the electrons' plane is parallel to **B**, therefore, the magnetic flux, Φ_B is always zero. This means that **F** and **E** are zero and as a consequence, there is no force on the electrons and there will be no Δv for the electrons. The $\omega_0 + \Delta\omega$ is the case calculated in parts (a)–(c). The $\omega_0 - \Delta\omega$ will have the same magnitude for **F**, **B**, and Δv as in (a)–(c) but the direction will be opposite.

3-3 (a) The total energy of a simple harmonic oscillator having an amplitude A is $\dfrac{kA^2}{2}$, therefore, $E = \dfrac{kA^2}{2} = (25 \ \text{N/m})\dfrac{(0.4 \ \text{m})^2}{2} = 2.0 \ \text{J}$. The frequency of oscillation will be $f = \left(\dfrac{1}{2\pi}\right)\left(\dfrac{k}{m}\right)^{1/2} = \left(\dfrac{1}{2\pi}\right)\left(\dfrac{25}{2}\right)^{1/2} = 0.56 \ \text{Hz}$.

(b) If energy is quantized, it will be given by $E_n = nhf$ and from the result of (a) there follows $E_n = nhf = n\left(6.63 \times 10^{-34} \ \text{J s}\right)(0.56 \ \text{Hz}) = 2.0 \ \text{J}$. Upon solving for n one obtains $n = 5.4 \times 10^{33}$.

(c) The energy carried away by one quantum of charge in energy will be $E = hf = \left(6.63 \times 10^{-34} \ \text{J s}\right)(0.56 \ \text{Hz}) = 3.7 \times 10^{-34} \ \text{J}$.

3-5 (a) Planck's radiation energy density law as a function of wavelength and temperature is given by $u(\lambda, T) = \dfrac{8\pi hc}{\lambda^5\left(e^{hc/\lambda_B T} - 1\right)}$. Using $\dfrac{\partial u}{\partial \lambda} = 0$ and setting $x = \dfrac{hc}{\lambda_{max} k_B T}$, yields an extremum in $u(\lambda, T)$ with respect to λ. The result is

$0 = -5 + \left(\dfrac{hc}{\lambda_{max} k_B T}\right)\left(e^{hc/\lambda_{max} k_B T}\right)\left(e^{hc/\lambda_{max} k_B T} - 1\right)^{-1}$ or $x = 5\left(1 - e^{-x}\right)$.

(b) Solving for x by successive approximations, gives $x \cong 4.965$ or $\lambda_{max} T = \left(\dfrac{hc}{k_B}\right)(4.965) = 2.90 \times 10^{-3} \ \text{m} \cdot \text{K}$.

3-7 (a) In general, $L = \dfrac{n\lambda}{2}$ where $n = 1, 2, 3, \ldots$ defines a mode or standing wave pattern with a given wavelength. As we wish to find the number of possible values of n between 2.0 and 2.1 cm, we use $n = \dfrac{2L}{\lambda}$

$$n(2.0 \ \text{cm}) = (2)\frac{200}{2.0} = 200$$
$$n(2.1 \ \text{cm}) = (2)\frac{200}{2.1} = 190$$
$$|\Delta n| = 10$$

As n changes by one for each allowed standing wave, there are 10 standing waves of different wavelength between 2.0 and 2.1 cm.

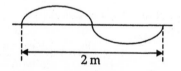

2 m

(b) The number of modes per unit wavelength per unit length is
$\dfrac{\Delta n}{L \Delta \lambda} = \dfrac{10}{0.1}(200) = 0.5 \ \text{cm}^{-2}$.

(c) For short wavelengths n is almost a continuous function of λ. Thus one may use calculus to approximate $\frac{\Delta n}{L\Delta\gamma} = \left(\frac{1}{L}\right)\left(\frac{dn}{d\lambda}\right)$. As $n = \frac{2L}{\lambda}$, $\left|\frac{dn}{d\lambda}\right| = \frac{2L}{\lambda^2}$ and $\left(\frac{1}{L}\right)\left(\frac{dn}{d\lambda}\right) = \frac{2}{\lambda^2}$. This gives approximately the same result as that found in (b):

$$\left(\frac{1}{L}\right)\left(\frac{dn}{d\lambda}\right) = \frac{2}{\lambda^2} = \frac{2}{(2.0 \text{ cm})^2} = 0.5 \text{ cm}^{-2}.$$

(d) For short wavelengths n is almost a continuous function of λ, $n = \frac{2L}{\lambda}$ is a discrete function.

3-9 Use $E = \frac{hc}{\lambda}$ or $\lambda = \frac{hc}{E}$ (where $hc = 1\,240$ eV nm) and the results of Problem 3-7 to find

(a) $\lambda = 600$ nm

(b) $\lambda = 0.03$ m

(c) $\lambda = 10$ m

3-11 Following the same reasoning as in Problem 3-9, one obtains

$$\frac{n}{t} = \frac{P}{hf} = \frac{P\lambda}{hc} = (3.74 \times 10^{26} \text{ J s}) \frac{500 \times 10^{-9} \text{ s}^{-1}}{6.63 \times 10^{-34} \text{ J s}} (3 \times 10^8 \text{ s}^{-1}) = 9.45 \times 10^{44} \text{ photons/s}.$$

3-13 $K = hf - \phi = \frac{hc}{\lambda} - \phi$

$$\phi = \frac{hc}{\lambda} - K = \frac{1\,240 \text{ eV nm}}{250 \text{ nm}} - 2.92 \text{ eV} = 2.04 \text{ eV}$$

3-15 (a) At the cut-off wavelength, $K = 0$ so $\frac{hc}{\lambda} - \phi = 0$, or $\lambda_{\text{cut-off}} = \frac{hc}{\phi} = \frac{1\,240 \text{ eV nm}}{4.2 \text{ eV}} = 300$ nm. The threshold frequency, f_0 is given by

$$f_0 = \frac{c}{\lambda_{\text{cut-off}}} = \frac{3.0 \times 10^8 \text{ m/s}}{3.0 \times 10^2 \times 10^{-9} \text{ m}} = 1.0 \times 10^{15} \text{ Hz}.$$

(b) $eV_s = K = hf - \phi = \frac{hc}{\lambda} - \phi$

$$V_s = \frac{hc}{\lambda e} - \frac{\phi}{e} = \frac{1\,240 \text{ eV nm}}{200 \text{ nm e}} = -4.2 \text{ eV/e} = 2.0 \text{ V}$$

3-17 The energy of one photon of light of wavelength $\lambda = 300$ nm is

$$E = \frac{hc}{\lambda} = \frac{1\,240 \text{ eV nm}}{300 \text{ nm}} = 4.13 \text{ eV}.$$

(a) As lithium and beryllium have work functions that are less than 4.13 eV, they will exhibit the photoelectric effect for incident light with this energy. However, mercury will not because its work function is greater than 4.13 eV.

(b)   The maximum kinetic energy is given by $K = \dfrac{hc}{\lambda - \phi}$, so

$$K(Li) = \frac{1\,240\text{ eV nm}}{300\text{ nm}} - 2.3\text{ eV} = 1.83\text{ eV, and } K(Be) = \frac{1\,240\text{ eV nm}}{300\text{ nm}} - 3.9\text{ eV} = 0.23\text{ eV}.$$

3-19 $\phi = 2.00\text{ eV}$, $K_{max} = eV_0 = hf - \phi = \dfrac{hc}{\lambda} - \phi$.

$$\Rightarrow V_0 = \frac{\left(\frac{hc}{\lambda} - \phi\right)}{e} = \frac{\frac{(4.14\times10^{-15}\text{ eV s})(3\times10^8\text{ m/s})}{350\times10^{-9}\text{ m}} - 2.00\text{ eV}}{e} = 1.55\text{ V}.$$

3-21 $V_s = \left(\dfrac{h}{e}\right)\dfrac{f - \phi}{e}$. Choosing two points on the graph, one has $\left(\dfrac{h}{e}\right)(4\times10^{14}\text{ Hz}) - \dfrac{\phi}{e} = 0$ and

$\left(\dfrac{h}{e}\right)(8\times10^{14}\text{ Hz}) - 1.7\text{ eV}$. Combining these two expressions one obtains:

(a) $\phi = 1.6\text{ eV}$

(b) $\dfrac{h}{e} = 4.0\times10^{-15}\text{ Vs}$

(c)   For cut-off wavelength, $\lambda_c = \dfrac{hc}{\phi} = \dfrac{1\,240\text{ eV nm}}{1.6\text{eV}} = 775\text{ nm}$.

(d)   Accepted $\dfrac{h}{e} = 4.14\times10^{-15}\text{ Vs}$, about a 3% difference.

3-23 $E = \dfrac{hc}{\lambda} = \dfrac{(6.626\times10^{-34}\text{ J s})(3\times10^8\text{ m/s})}{(5\times10^{-7}\text{ m})(1.6\times10^{-19}\text{ J/eV})} = 2.48\text{ eV}$

$p = \dfrac{h}{\lambda} = \dfrac{E}{c} = \dfrac{(2.48\text{ eV})(1.6\times10^{-19}\text{ J/eV})}{3\times10^8\text{ m/s}} = 1.32\times10^{-27}\text{ kg m/s}$

3-25 $E = 300\text{ keV}$, $\theta = 30°$

(a) $\Delta\lambda = \lambda' - \lambda_0 = \dfrac{h}{m_e c}(1 - \cos\theta) = (0.002\,43\text{ nm})[1 - \cos(30°)] = 3.25\times10^{-13}\text{ m}$

$= 3.25\times10^{-4}\text{ nm}$

(b) $E = \dfrac{hc}{\lambda_0} \Rightarrow \lambda_0 = \dfrac{hc}{E_0} = \dfrac{(4.14\times10^{-15}\text{ eVs})(3\times10^8\text{ m/s})}{300\times10^3\text{ eV}} = 4.14\times10^{-12}\text{ m}$; thus,

$\lambda' = \lambda_0 + \Delta\lambda = 4.14\times10^{-12}\text{ m} + 0.325\times10^{-12}\text{ m} = 4.465\times10^{-12}\text{ m}$, and

$E' = \dfrac{hc}{\lambda'} \Rightarrow E' = \dfrac{(4.14\times10^{-15}\text{ eV s})(3\times10^8\text{ m/s})}{4.465\times10^{-12}\text{ m}} = 2.78\times10^5\text{ eV}$.

(c) $\dfrac{hc}{\lambda_0} = \dfrac{hc}{\lambda'} + K_e$, (conservation of energy)

$$K_e = hc\left(\frac{1}{\lambda_0} - \frac{1}{\lambda'}\right) = \frac{(4.14\times10^{-15}\text{ eV s})(3\times10^8\text{ m/s})}{\frac{1}{4.14\times10^{-12}} - \frac{1}{4.465\times10^{-12}}} = 22\text{ keV}$$

3-27   Conservation of energy yields $hf = K_e + hf'$ (Equation A). Conservation of momentum yields $p_e^2 = p'2 + p^2 - 2pp'\cos\theta$. Using $p_{photon} = \dfrac{E}{c} = \dfrac{hf}{c}$ there results

$p_e^2 = \left(\dfrac{hf'}{c}\right)^2 + \left(\dfrac{hf}{c}\right)^2 - 2\left(\dfrac{hf}{c}\right)\left(\dfrac{hf'}{c}\right)\cos\theta$ (Equation B). If the photon transfers all of its energy,

$f' = 0$ and Equations A and B become $K_e = hf$ and $p_e^2\left(\dfrac{hf}{c}\right)^2$ respectively. Note that in general,

$K_e = E_e - m_e c^2 = \left[p_e^2 c^2 + \left(m_e c^2\right)^2\right]^{1/2} - m_e c^2$. Finally, substituting $K_e = hf$ and $P_e^2 = \left(\dfrac{hf}{c}\right)^2$ into

$K_e = \left[p_e^2 c^2 + \left(m_e c^2\right)^2\right]^{1/2} - m_e c^2$, yields $hf = \left[(hf)^2 + \left(m_e c^2\right)^2\right]^{1/2} - m_e c^2$ (Equation C). As Equation C is true only if h, or f, or m_e, or c is zero and all are non-zero this contradiction means that f' cannot equal zero and conserve both relativistic energy and momentum.

3-29   Symmetric Scattering, $\theta = \phi$. First, use the equations of conservations of momentum given by Equations 3.30 and 3.31 for this two dimensional scattering process with $\theta = \phi$:

(a) $\dfrac{h}{\lambda_0} = \left(\dfrac{h}{\lambda'}\right)\cos\theta + p_e\cos\theta$ (1)

$\dfrac{h}{\lambda'}\sin\theta = p_e\sin\theta$ or $p_e = \dfrac{h}{\lambda}$ (2)

Substituting (2) into (1) yields $\lambda' = 2\lambda_0\cos\theta$ (3)

Next, express the Compton scattering formula as

$\lambda' - \lambda_0 = \lambda_c(1 - \cos\theta)$ (4)

where $\lambda_c = \dfrac{h}{m_e c} = 0.002\,43$ nm. Combining (3) and (4) yields $\cos\theta = \dfrac{\lambda_c + \lambda_0}{\lambda_c + 2\lambda_0}$. In this case, because $E = 1.02$ MeV, and $E = \dfrac{hc}{\lambda_0}$ there results

$$\lambda_0 = \frac{hc}{E} = \frac{1\,240\text{ eV nm}}{1.20 \times 106\text{ eV}} = 0.001\,22\text{ nm}.$$

Thus, $\cos\theta = \dfrac{0.002\,43\text{ nm} + 0.001\,22\text{ nm}}{0.002\,42\text{ nm} + 0.002\,44\text{ nm}} = 0.749\,5$, and solving for the scattering angle, $\theta = 41.5°$.

(b) $\lambda' = \lambda_0 + \lambda_c(1 - \cos\theta)$
$\lambda' = 0.001\,22\text{ nm} + (0.002\,43\text{ nm})[1 - \cos(41.5°)] = 0.001\,83\text{ nm}$

$E = \dfrac{hc}{\lambda'} = \dfrac{1\,240\text{ eV}\cdot\text{nm}}{0.001\,83\text{ nm}} = 0.679\text{ MeV}$

3-31 (a) $E' = \dfrac{hc}{\lambda'}$, $\lambda' = \lambda_0 + \Delta\lambda$

$$\lambda_0 = \frac{hc}{E_0} = \frac{(6.63 \times 10^{-34} \ \text{J} \cdot \text{s})(3 \times 10^8 \ \text{m/s})}{0.1 \ \text{MeV}} = 1.243 \times 10^{-11} \ \text{m}$$

$$\Delta\lambda = \left(\frac{h}{m_e c}\right)(1 - \cos\theta) = \frac{(6.63 \times 10^{-34} \ \text{J} \cdot \text{s})(1 - \cos 60°)}{(9.11 \times 10^{-34} \ \text{kg})(3 \times 10^8 \ \text{m/s})} = 1.215 \times 10^{-12} \ \text{m}$$

$$\lambda' = \lambda_0 + \Delta\lambda = 1.364 \times 10^{-11} \ \text{m}$$

$$E' = \frac{hc}{\lambda'} = \frac{(6.63 \times 10^{-34} \ \text{J} \cdot \text{s})(3 \times 10^8 \ \text{m/s})}{1.364 \times 10^{-11} \ \text{m}} = 9.11 \times 10^4 \ \text{eV}$$

(b) $\dfrac{hc}{\lambda_0} = \dfrac{hc}{\lambda'} + K_e$

$K_e = 0.1 \ \text{MeV} - 91.1 \ \text{keV} = 8.90 \ \text{keV}$

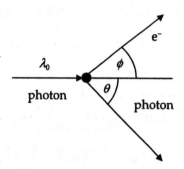

(c)   Conservation of momentum along x: $\dfrac{h}{\lambda_0} = \left(\dfrac{h}{\lambda'}\right)\cos\theta + \gamma m_e v \cos\phi$. Conservation of momentum along y: $\left(\dfrac{h}{\lambda'}\right)\sin\theta = \gamma m_e v \sin\phi$. So that

$$\frac{\gamma m_e v \sin\phi}{\gamma m_e v \cos\phi} = \left(\frac{h}{\lambda'}\right)\sin\theta\left[\left(\frac{h}{\lambda_0}\right) - \left(\frac{h}{\lambda'}\right)\cos\theta\right]$$

$$\tan\phi = \frac{\lambda_0 \sin\theta}{(\lambda' - \lambda_0)\cos\theta}$$

Here, $\theta = 60°$, $\lambda_0 = 1.243 \times 10^{-11}$ m, and $\lambda' = 1.364 \times 10^{-11}$ m. Consequently,

$$\tan\phi = \frac{(1.24 \times 10^{-11} \ \text{m})(\sin 60°)}{(1.36 - 1.24 \cos 60°) \times 10^{-11} \ \text{m}} = 1.451$$

$$\phi = 55.4°$$

3-33   Substituting equations 3-33 and 3-34 of the text, $E_e = h(f_0 - f') + m_e c^2$ and

$$p_e^2 c^2 = h^2\left(f'^2 + f_0^2\right) - 2h^2 f' f_0 \cos\theta$$

into the relativistic energy expression $E_e^2 = p_e^2 c^2 + \left(m_e c^2\right)^2$ yields

$$h^2\left(f'^2 + f_0^2 - 2f_0 f'\right) + m_e^2 c^4 + 2h(f_0 - f') m_e c^2 = h^2\left(f'^2 + f_0^2\right) - 2h^2 f_0 f' \cos\theta^2 + \left(m_e c^2\right)^2.$$

Canceling and combining there results

$$\left(f'^2 + f_0^2 - 2f_0 f'\right) + \frac{2 m_e c^2 (f_0 - f')}{h} = f'^2 + f_0^2 - 2f_0 f' \cos\theta$$

which reduces to $\dfrac{m_e c^2 (f_0 - f')}{h} = f_0 f'(1 - \cos\theta)$. Using $\lambda f = c$ one obtains

$\lambda' - \lambda_0 = \dfrac{h(1 - \cos\theta)}{m_e c}$, which is the Compton scattering or Compton shift relation.

3-35   (a)   The energy vs wavelength relation for a photon is $E = \dfrac{hc}{\lambda}$. For a photon of wavelength given by $\lambda_0 = 0.0711$ nm the photon's energy is

$$E = \frac{\left(6.626 \times 10^{-34}\ \text{J} \cdot \text{s}\right)\left(3 \times 10^8\ \text{m/s}\right)}{\left(0.0711 \times 10^{-9}\ \text{m}\right)\left(1.602 \times 10^{-19}\ \text{J/eV}\right)} = 17.4\ \text{keV}$$

(b)   For the case of back scattering, $\theta = \pi$ the Compton scattering relation becomes $\lambda' - \lambda_0 = \left(\dfrac{2hc}{m_e c^2}\right)$. Setting $\lambda_0 = 0.0711$ nm we obtain

$$\lambda' = 0.711\ \text{nm} + \frac{2hc}{m_e c^2} = 7.60 \times 10^{-11}$$

or 0.0760 nm.

(c) $E' = \dfrac{hc}{\lambda'} = \dfrac{\left(6.626 \times 10^{-34}\ \text{J} \cdot \text{s}\right)\left(3 \times 10^8\ \text{m/s}\right)}{\left(7.60 \times 10^{-11}\ \text{m}\right)\left(1.602 \times 10^{-19}\ \text{J/eV}\right)} = 16.3\ \text{keV}.$

(d) $\Delta E = 17.45\ \text{keV} - 16.33\ \text{keV} = 1.12\ \text{keV} \sim 1.1\ \text{keV}.$

3-37 When waves are scattered between two adjacent planes of a single crystal, constructive wave interference will occur when the path length difference between such reflected waves is an integer multiple of wavelengths. This condition is expressed by the Bragg equation for constructive interference, $2d \sin\theta = n\lambda$ where d is the distance between adjacent crystalline planes, θ is the angle of incidence of the x-ray beam of photons, n is an integer for constructive interference, and λ is the wavelength of the photon beam which is in this case, 0.062 6 nm. Ignoring the incident beam that is not scattered, the first three angles for which maxima of x-ray intensities are found are $1\lambda = 2d \sin\theta_1$ or

$$\sin\theta_1 = \frac{\lambda}{2d} = \frac{0.626 \times 10^{-10} \text{ m}}{8 \times 10^{-10} \text{ m}}$$
$$\theta_1 = 0.078\ 3 \text{ radians} = 4.49°$$

$2\lambda = 2d \sin\theta_2$ or

$$\sin\theta_2 = \frac{\lambda}{d} = \frac{0.626 \times 10^{-10} \text{ m}}{4.0 \times 10^{-10} \text{ m}} = 0.156\ 5, \ \theta = 9.00°$$

$3\lambda = 2d \sin\theta_3$ or

$$\sin\theta_3 = \frac{3\lambda}{2d} = \frac{3(0.626 \times 10^{-10} \text{ m})}{8 \times 10^{-10} \text{ m}} = 0.234\ 75, \ \theta_3 = 13.6°$$

3-39 The first x-ray intensity maximum in the diffraction pattern occurs at $\theta = 6.41°$. To determine d use the Bragg diffraction condition $n\lambda = 2d \sin\theta$ for $n = 1$.

$$d = \frac{\lambda}{2\sin\theta} = \frac{0.626 \text{ Å}}{2\sin 6.41°} = 2.80 \text{ Å} = 2.80 \times 10^{-8} \text{ cm}.$$

From Figure P3.39 there are $(4)\left(\frac{1}{8}\right)Cl^-$ and $(4)\left(\frac{1}{8}\right)Na^+$ ions per primitive cell. This works out to half a NaCl formula unit per primitive cell. The formula weight of NaCl is 58.4 g/mole. Setting the mass per unit volume of the primitive cell equal to the density we have $\frac{(58.4 \text{ g/mole})(\text{formula unit})}{2d^3 N_A} = \rho$ where N_A is the number of formula units per mole (Avogadro's number). So

$$N_A = \frac{(58.4 \text{ g/mol})(\text{formula unit})}{2d^3}$$
$$\rho = \frac{(58.4 \text{ g/mol})(\text{formula unit})}{2(2.8 \times 10^{-8} \text{ cm})^3 (2.17 \text{ g/cm}^3)}$$
$$N_A = 6.13 \times 10^{23} \text{ formula units/mole}$$

3-41 The classical definition of a black hole is a star so massive that no object nor any electromagnetic radiation can escape its gravitational attraction. This occurs when the star's mass is such that the gravitational term $\left(\frac{GM_S}{c^2 R_S}\right)$ is greater than unity. Setting this term to be unity and using M_S to be the mass of the sun yields a value of

(a) $R_{\text{blackhole}} \approx \dfrac{GM_S}{c^2} = \dfrac{\left(6.67 \times 10^{-11} \ \text{N} \cdot \text{m}^2/\text{kg}^2\right)\left(1.99 \times 10^{30}\right)}{\left(3.00 \times 10^8 \ \text{m/s}\right)^2} = 1\,470 \ \text{m} \approx 1\,500 \ \text{m}$ for a *black*

hole of one solar mass. The normal sun has a radius of 6.96×10^8 m.

(b)   The density of the black hole is $\rho_{\text{blackhole}} = \dfrac{M_S}{4\pi(R_{\text{blackhole}})^3/3}$ and the density of the

sun is $\rho_S = \dfrac{M_S}{4\pi(R_S)^3/3}$. Then

$$\frac{\rho_{\text{blackhole}}}{\rho_S} = \frac{3M_S/4\pi(R_{\text{blackhole}})^3}{\rho_{\text{blackhole}}} = \frac{3M_S}{4\pi(R_S)^3} = \frac{R_S^3}{(R_{\text{blackhole}})^3} = \frac{\left(6.96 \times 10^8 \ \text{m}\right)^3}{\left(1.47 \times 10^3 \ \text{m}\right)^3} = 1.1 \times 10^{17}$$

or an increase in density over that of the sun by the factor of a *wicked* 10^{17} or seventeen orders of magnitude.

3-43   (a)   A 4000 Å wavelength photon is backscattered, $\theta = \pi$ by an electron. The energy transferred to the electron is determined by using the Compton scattering formula

$\lambda' - \lambda_0 = \left(\dfrac{hc}{E_e}\right)(1 - \cos\theta)$ where we take $E_e = m_e c^2$ for the rest energy of the electron

and so $E_e \approx 0.511$ MeV. Upon substitution, one obtains

$$\Delta\lambda = 2(0.002\,43 \ \text{nm}) = 0.004\,86 \ \text{nm}.$$

The energy of a photon is related to its wavelength by the relation $E = \dfrac{hc}{\lambda}$, so the

change in energy associated with a corresponding change in wavelength is given by $\Delta E = -\left(\dfrac{hc}{\lambda^2}\right)\Delta\lambda$. Upon making substitutions one obtains the magnitude

$\Delta E = 6.037\,9 \times 10^{-24}$ J and using the conversion factor 1 Joule of energy is equivalent to 1.602×10^{-19} eV. The result is $\Delta E = 3.77 \times 10^{-5}$ eV.

(b)   This may be compared to the energy that would be acquired by an electron in the photoelectric effect process. Here again the energy of a photon of wavelength λ is

given by $E = \dfrac{hc}{\lambda}$. With $\lambda = 400$ nm, one obtains

$$E = \frac{\left(6.626 \times 10^{-34} \ \text{J} \cdot \text{s}\right)\left(3.0 \times 10^8 \ \text{m/s}\right)}{400 \times 10^{-9} \ \text{m}} = 4.97 \times 10^{-19} \ \text{J}$$

and upon converting to electron volts, $E = 3.10$ eV . $\dfrac{\Delta E}{E_{\text{photon}}} \approx 10^{-5}$. The maximum

energy transfer is about five orders of magnitude smaller than the energy necessary for the photoelectric effect.

(c)   Could "a violet photon" eject an electron from a metal by Compton scattering? The answer is no, because the maximum energy transfer occurring at $\theta = \pi$ is not sufficient.

3-45   From a plot of V_s versus f one finds $h = 6.7 \times 10^{-34}$ J·s, $f_0 = 7.1 \times 10^{14}$ Hz, and $\phi = 2.29$ eV.

3-47   Head on collision means the point of contact is through the center of mass of the electron, therefore, it is a case of back scattering. Again, using the Compton shift formula

$$\Delta \lambda = h \frac{1 - \cos 180°}{m_e c} = \frac{2h}{m_e c}$$

$$E_0 = \frac{hc}{\lambda_0}$$

$$\lambda_f = \lambda_0 + \Delta \lambda = \frac{hc}{E_0} + \frac{2h}{m_e c}$$

$$E_{electron} = E_0 = \frac{hc}{\lambda_0 + \Delta \lambda} = E_0 - \frac{hc}{\frac{hc}{E_0} + \frac{2h}{m_e c}} = \frac{\left(\frac{2h}{m_e c}\right) E_0}{\frac{hc}{E_0} + \frac{2h}{m_e c}} = \frac{\frac{2E_0 h}{m_e c}}{\left(\frac{hc}{E_0}\right)\left[1 + \left(\frac{2E_0}{m_e c^2}\right)\right]}$$

$$E_{electron} = \left(\frac{2E_0^2}{m_e c^2}\right)\left[1 + \left(\frac{2E_0}{mc^2}\right)^{-1}\right] = 2hf\alpha(1 + 2\alpha)^{-1} \text{ where } \alpha = \frac{E_0}{m_e c^2}.$$

photon   electron

4

The Particle Nature of Matter

4-1 F corresponds to the charge passed to deposit one mole of monovalent element at a cathode. As one mole contains Avogadro's number of atoms, $e = \dfrac{96\,500\ \text{C}}{6.02 \times 10^{23}} = 1.60 \times 10^{-19}\ \text{C}$.

4-3 Thomson's device will work for positive and negative particles, so we may apply $\dfrac{q}{m} \approx \dfrac{V\theta}{B^2 ld}$.

(a) $\dfrac{q}{m} \approx \dfrac{V\theta}{B^2 ld} = (2\,000\ \text{V}) \dfrac{0.20\ \text{radians}}{\left(4.57 \times 10^{-2}\ \text{T}\right)^2} (0.10\ \text{m})(0.02\ \text{m}) = 9.58 \times 10^7\ \text{C/kg}$

(b) As the particle is attracted by the negative plate, it carries a positive charge and is a proton. $\left(\dfrac{q}{m_p} = \dfrac{1.60 \times 10^{-19}\ \text{C}}{1.67 \times 10^{-27}\ \text{kg}} = 9.58 \times 10^7\ \text{C/kg} \right)$

(c) $v_x = \dfrac{E}{B} = \dfrac{V}{dB} = \dfrac{2\,000\ \text{V}}{0.02\ \text{m}} \left(4.57 \times 10^{-2}\ \text{T} \right) = 2.19 \times 10^6\ \text{m/s}$

(d) As $v_x \sim 0.01c$ there is no need for relativistic mechanics.

4-5 (a) Draw the curved path of the electron as follows: $(r - y)^2 + l^2 = r^2$

$$r^2 - 2ry + y^2 + l^2 = r^2$$

$$r = \dfrac{l^2 + y^2}{2y}$$

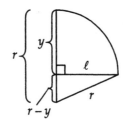

(b) $v = \dfrac{V}{Bd} = \dfrac{1\,060\text{ V}}{0.017\,7}(2.5\times10^{-4}\text{ m}) = 2.39\times10^8\text{ m/s} = 0.795c$. Using the classical expression

$p = mv = Bre$ gives $r = \dfrac{mv}{Be} = (9.11\times10^{-31}\text{ kg})\dfrac{2.39\times10^8\text{ m/s}}{0.017\,7\text{ T}}(1.60\times10^{-19}\text{ C}) = 0.076\,9\text{ m}$.

To find the value of y corresponding to this r use $y^2 - 2ry + l^2 = 0$ or

$y^2 - 0.153\,8y + 6.10\times10^{-4} = 0$. Using the quadratic formula,

$$y = -b \pm \dfrac{\left(b^2 - 4ac\right)^{1/2}}{2a}$$

$$y = +0.153\,8 \pm \dfrac{\left[(0.153\,8)^2 - (4)\left(6.101\times10^{-4}\right)\right]^{1/2}}{2} = 0.150\text{ m}$$

$4.08\times10^{-3}\text{ m} = 0.004\,08\text{ m}$ (the first answer is discarded because $r > y$). Using the relativistic expression $p = \gamma mv = Bre$ gives

$$r = \dfrac{\gamma me}{Be} = \dfrac{(1)\left(9.11\times10^{-31}\text{ kg}\right)\left(2.39\times10^8\text{ m/s}\right)}{\left[1 - (0.795)^2\right]^{1/2}(0.017\,7\text{ T})(1.60\times10^{-19}\text{ C})} = 0.126\,7\text{ m}.$$

The y values corresponding to this r value are found from $y^2 - 0.253y + 6.10\times10^{-4} = 0$ or $y = 0.251\text{ m},\ 0.002\,43\text{ m}$. Discarding the first answer because r must be greater than y, leaves $y = 0.002\,43\text{ m}$ in good agreement with $y_{\text{observed}} = (0.002\,4 \pm 0.000\,5)\text{m}$.

4-7 $m = \rho V = (0.956\,1\text{ g/cm}^3)\left(\dfrac{4}{3}\right)\pi a^3 = 8.418\times10^{-11}\text{ g}$, $\dfrac{mg}{E}$ for use in $q = \left(\dfrac{mg}{E}\right)\dfrac{v + v_1'}{v}$ has the

value $\dfrac{mg}{E} = (8.418\times10^{-14}\text{ kg})\dfrac{9.80\text{ m/s}^2}{\frac{5\,085\text{ V}}{0.016\,00\text{ m}}} = 25.9\times10^{-19}\text{ C}$. Using $q = \left(\dfrac{mg}{E}\right)\dfrac{v + v_1'}{v}$ we find the

different charges on the drops to be:

$$q_1 = \left(25.96\times10^{-19}\text{ C}\right)\dfrac{0.858\,42 + 0.126\,5}{0.858\,42} = 29.78\times10^{-19}\text{ C}$$

$$q_2 = 39.76\times10^{-19}\text{ C}$$

$$q_3 = 28.16\times10^{-19}\text{ C}$$

$$q_4 = 29.84\times10^{-19}\text{ C}$$

$$q_5 = 34.84\times10^{-19}\text{ C}$$

$$q_6 = 36.51\times10^{-19}\text{ C}$$

$[|\text{Charge differences}|] \times 10^{-19}$ C	$\left[\dfrac{|\text{Charge differences}|}{n}\right] \times 10^{-19}$ C (n chosen by inspection)
$q_1 - q_2 = 9.98$	$\dfrac{9.98}{6} = 1.66$
$q_1 - q_3 = 1.62$	$\dfrac{1.62}{1} = 1.62$
$q_1 - q_4 = 0.060$	0.0
$q_1 - q_5 = 5.06$	$\dfrac{5.06}{3} = 1.69$
$q_1 - q_6 = 6.73$	$\dfrac{6.73}{4} = 1.68$
$q_3 - q_2 = 11.6$	$\dfrac{11.6}{7} = 1.66$
$q_4 - q_2 = 9.92$	$\dfrac{9.92}{6} = 1.65$
$q_5 - q_2 = 4.92$	$\dfrac{4.92}{3} = 1.64$
$q_6 - q_2 = 3.25$	$\dfrac{3.25}{2} = 1.63$
$q_4 - q_3 = 1.68$	$\dfrac{1.68}{2} = 1.68$
$q_5 - q_3 = 6.68$	$\dfrac{6.68}{4} = 1.67$
$q_6 - q_3 = 8.35$	$\dfrac{8.35}{5} = 1.67$
$q_5 - q_4 = 5.00$	$\dfrac{5.00}{3} = 1.67$
$q_6 - q_4 = 6.67$	$\dfrac{6.67}{4} = 1.66$
$q_6 - q_5 = 1.67$	$\dfrac{1.67}{1} = 1.67$

Average $q = 1.661 \times 10^{-19}$ C

4-9   The initial energy of the system of α plus copper nucleus is 13.9 MeV and is just the kinetic energy of the α when the α is far from the nucleus. The final energy of the system may be evaluated at the point of closest approach when the kinetic energy is zero and the potential energy is $k(2e)\dfrac{Ze}{r}$ where r is approximately equal to the nuclear radius of copper. Invoking conservation of energy $E_i = E_f$, $K_\alpha = (k)\dfrac{2Ze^2}{r}$ or

$$r = (k)\frac{2Ze^2}{K_\alpha} = \frac{(2)(29)\left(1.60 \times 10^{-19}\right)^2 \left(8.99 \times 10^9\right)}{\left(13.9 \times 10^6 \text{ eV}\right)\left(1.60 \times 10^{-19} \text{ J/eV}\right)} = 6.00 \times 10^{-15} \text{ m.}$$

4-11 $\dfrac{1}{\lambda} = R\left(\dfrac{1}{n_f^2} - \dfrac{1}{n_i^2}\right)$. For the Balmer series, $n_f = 2$; $n_i = 3, 4, 5, \ldots$. The first three lines in the series have wavelengths given by $\dfrac{1}{\lambda} = R\left(\dfrac{1}{2^2} - \dfrac{1}{n^2}\right)$ where $R = 1.097\,37 \times 10^7 \text{ m}^{-1}$.

$$\text{1st line: } \frac{1}{\lambda} = R\left(\frac{1}{4} - \frac{1}{9}\right) = \left(\frac{5}{36}\right)R; \ \lambda = \frac{36}{5R} = 656.112 \text{ nm}$$

$$\text{2nd line: } \frac{1}{\lambda} = R\left(\frac{1}{4} - \frac{1}{16}\right) = \left(\frac{3}{16}\right)R; \ \lambda = \frac{16}{3R} = 486.009 \text{ nm}$$

$$\text{3rd line: } \frac{1}{\lambda} = R\left(\frac{1}{4} - \frac{1}{25}\right) = \left(\frac{21}{100}\right)R; \ \lambda = \frac{100}{21R} = 433.937 \text{ nm}$$

4-13   (a) $\lambda = 102.6 \text{ nm}; \ \dfrac{1}{\lambda} = R\left(1 - \dfrac{1}{n^2}\right) \Rightarrow n = \dfrac{R}{\left(R - \frac{1}{\lambda}\right)^{1/2}} = \dfrac{R}{\left(R - \frac{1}{102.6 \times 10^{-9} \text{ m}}\right)^{1/2}} = 2.99 \approx 3$

       (b)   This wavelength cannot belong to either series. Both the Paschen and Brackett series lie in the IR region, whereas the wavelength of 102.6 nm lies in the UV region.

4-15   (a)   The energy levels of a hydrogen-like ion whose charge number is 2 is given by

$$E_n = (-13.6 \text{ eV})\frac{Z^2}{n^2} = \frac{-54.4 \text{ eV}}{n^2} \text{ for } Z = 2. \ \left(\text{He}^+\right)$$

$$\underline{\hspace{6cm}} \quad 0$$
$$\underline{\hspace{6cm}} \quad E_3 = -6.04 \text{ eV}$$
$$\underline{\hspace{6cm}} \quad E_2 = -13.6 \text{ eV}$$

So $E_1 = -54.4 \text{ eV}$
       $E_2 = -13.6 \text{ eV}$
       $E_3 = -6.04 \text{ eV}$, etc.

$$\underline{\hspace{6cm}} \quad E_1 = -54.4 \text{ eV}$$

       (b)   For He^+, $Z = 2$ so we see that the ionization energy (the energy required to take the electron from the state $n = 1$ to the state $n = \infty$ is $E = (-13.6 \text{ eV})\dfrac{2^2}{1^2} = \dfrac{-54.4 \text{ eV}}{n^2}$ for $Z = 2$. $\left(\text{He}^+\right)$

4-17 $r = \dfrac{n^2 \hbar^2}{Z m_e k e^2} = \left(\dfrac{n^2}{Z}\right)\left(\dfrac{\hbar^2}{m_e k e^2}\right); \; n = 1$

$$r = \dfrac{1}{Z}\left[\dfrac{\left(1.055 \times 10^{-34} \text{ Js}\right)^2}{\left(9.11 \times 10^{-31} \text{ kg}\right)\left(9 \times 10^9 \text{ Nm}^2/\text{C}^2\right)\left(1.6 \times 10^{-19} \text{ C}\right)^2}\right] = \dfrac{5.30 \times 10^{-11} \text{ m}}{Z}$$

(a)   For He^+, $Z = 2$, $r = \dfrac{5.30 \times 10^{-11} \text{ m}}{2} = 2.65 \times 10^{-11} \text{ m} = 0.026\,5 \text{ nm}$

(b)   For Li^{2+}, $Z = 3$, $r = \dfrac{5.30 \times 10^{-11} \text{ m}}{3} = 1.77 \times 10^{-11} \text{ m} = 0.017\,7 \text{ nm}$

(c)   For Be^{3+}, $Z = 4$, $r = \dfrac{5.30 \times 10^{-11} \text{ m}}{4} = 1.32 \times 10^{-11} \text{ m} = 0.013\,2 \text{ nm}$

4-19   (a) $\Delta E = (-13.6 \text{ eV})\left(\dfrac{1}{n_i^2} - \dfrac{1}{n_f^2}\right) = (-13.6 \text{ eV})\left(\dfrac{1}{9} - \dfrac{1}{4}\right) = 1.89 \text{ eV}$

(b) $E = \dfrac{hc}{\lambda} \Rightarrow \lambda = \dfrac{hc}{E} = \left(4.14 \times 10^{-15} \text{ eV s}\right)\dfrac{3 \times 10^8 \text{ m/s}}{1.89 \text{ eV}} = 658 \text{ nm}$

(c) $c = \lambda f \Rightarrow f = \dfrac{c}{\lambda} = \dfrac{3 \times 10^8 \text{ m/s}}{657 \times 10^{-9} \text{ m}} = 4.56 \times 10^{14} \text{ Hz}$

4-21   (a)   For the Paschen series; $\dfrac{1}{\lambda} = R\left(\dfrac{1}{3^2} - \dfrac{1}{n_i^2}\right)$; the maximum wavelength corresponds to

$n_i = 4$, $\dfrac{1}{\lambda_{max}} = R\left(\dfrac{1}{3^2} - \dfrac{1}{4^2}\right)$; $\lambda_{max} = 1\,874.606 \text{ nm}$. For minimum wavelength, $n_i \to \infty$,

$\dfrac{1}{\lambda_{min}} = R\left(\dfrac{1}{3^2} - \dfrac{1}{\infty}\right)$; $\lambda_{min} = \dfrac{9}{R} = 820.140 \text{ nm}$.

(b) $\dfrac{hc}{\lambda_{min}} = \dfrac{\left(\frac{hc}{1\,874.606 \text{ nm}}\right)}{1.6 \times 10^{-19} \text{ J/eV}} = 0.662\,7 \text{ nm}$

$\dfrac{hc}{\lambda_{min}} = \dfrac{\left(\frac{hc}{820.140 \text{ nm}}\right)}{1.6 \times 10^{-19} \text{ J/eV}} = 1.515 \text{ nm}$

4-23   (a) $r_1 = (0.052\,9 \text{ nm})n^2 = 0.052\,9 \text{ nm}$ (when $n = 1$)

(b) $m_e v = m_e \left(\dfrac{k e^2}{m_e r}\right)^{1/2}$

$m_e = \left[\dfrac{\left(9.1 \times 10^{-31} \text{ kg}\right)\left(9 \times 10^9 \text{ Nm}^2/\text{C}^2\right)}{5.29 \times 10^{-11} \text{ m}}\right]^{1/2} \times \left(1.6 \times 10^{-19} \text{ C}\right)$

$M_e v = 1.99 \times 10^{-24} \text{ kg m/s}$

(c) $L = m_e v r = \left(1.99 \times 10^{-24} \text{ kg m/s}\right)\left(5.29 \times 10^{-11} \text{ m}\right)$, $L = 1.05 \times 10^{-34}\left(\text{kg m}^2/\text{s}\right) = \hbar$

(d) $K = |E| = 13.6$ eV

(e) $U = -2K = -27.2$ eV

(f) $E = K + U = -13.6$ eV

4-25 (a) $\Delta E = hf = (13.6 \text{ eV})\left(\dfrac{1}{n_f^2} - \dfrac{1}{n_i^2}\right)$ or $f = (13.6 \text{ eV})\left(\dfrac{\frac{1}{9} - \frac{1}{16}}{4.14 \times 10^{-15} \text{ eV s}}\right) = 1.60 \times 10^{14}$ Hz

(b) $T = \dfrac{2\pi r_n}{v}$ so $f_{rev} = \dfrac{1}{T} = \dfrac{v}{2\pi r_n}$. Using $v = \left(\dfrac{ke^2}{m_e r_n}\right)^{1/2}$, $f_{rev} = \left(\dfrac{ke^2}{m r_n}\right)^{1/2}$. For $n = 3$,

$r_3 = (3)^2 a_0$ and

$$f_{rev} = \frac{(8.99 \times 10^9 \text{ Nm}^2/\text{C}^2)(1.60 \times 10^{-19} \text{ C})^2}{\dfrac{[(9.11 \times 10^{-31} \text{ kg})(9)(5.29 \times 10^{-11} \text{ m})]^{1/2}}{(2)(3.14)(9)(5.29 \times 10^{-11} \text{ m})}}$$

$$f_{rev} = 2.44 \times 10^{14} \text{ Hz } (n = 3)$$

$$f_{rev} = 1.03 \times 10^{14} \text{ Hz } (n = 4)$$

Thus the photon frequency is about halfway between the two frequencies of the revolution.

4-27 (a) $\lambda = \dfrac{C_2 n^2}{n^2 - 2^2}$ so $\dfrac{1}{\lambda} = \left(\dfrac{1}{C_2}\right)\dfrac{n^2 - 2^2}{n^2} = \left(\dfrac{1}{C_2}\right)\left(\dfrac{1 - 2^2}{n^2}\right)$ or $\dfrac{1}{\lambda} = \left(\dfrac{2^2}{C_2}\right)\left(\dfrac{1}{2^2} - \dfrac{1}{n^2}\right) = R\left(\dfrac{1}{2^2} - \dfrac{1}{n^2}\right)$

where $R = \dfrac{2^2}{C_2}$.

(b) $R = \dfrac{2^2}{36\,545.6 \times 10^{-8} \text{ cm}} = 109\,720 \text{ cm}^{-1}$

4-29 (a) Both momentum and energy must be conserved. Momentum: $Mv = \dfrac{E_{photon}}{c}$, energy:

$E = E_{photon} + \dfrac{1}{2}Mv^2$ where M is the atom's mass, v is its recoil velocity, E_{photon} is the photon's energy and E is the energy difference between the $n = 3$ and $n = 1$ states.

Combining equations, $v^2 + 2cv - \dfrac{2E}{M} = 0$ and using the quadratic formula,

$$v = \frac{-2c \pm \left(\frac{4c^2 + 8E}{M}\right)^{1/2}}{2} = c\left[-1 \pm \left(\frac{1 + 2E}{Mc^2}\right)^{1/2}\right].$$

For $\dfrac{2E}{Mc^2} \ll 1$, $v \cong \dfrac{E}{Mc}$ or $v \cong \dfrac{-2c - E}{Mc}$, the second of which is non physical. Thus in

general, $v \cong \dfrac{E}{Mc}$. For the $n = 3$ to $n = 1$ transition in particular,

$$E = \left(\frac{ke^2}{2a_0}\right)\left(\frac{1}{n_f^2} - 1n_i^2\right) = hcR\left(1 - \frac{1}{9}\right) = \frac{8hcR}{9},$$

so $v \cong \dfrac{E}{Mc} = \dfrac{8hcR}{9Mc} = \dfrac{8hR}{9M}$.

(b)   The percent carried off $= \dfrac{\left(\frac{1}{2}Mv^2\right)}{\frac{8hcR}{9}}(100) = \dfrac{400hR}{9Mc}$

$$= \frac{(400)(6.63 \times 10^{-34}\ \text{Js})(1.10 \times 10^7\ \text{m}^{-1})}{(9)(93.34 \times 10^{-27}\ \text{kg})(3.00 \times 10^8\ \text{m/s})} = 3.23 \times 10^{-8}$$

4-31   (a) $(m_e vr) = n\hbar$; $v = \dfrac{n\hbar}{m_e r}$,

$$f_e = \frac{v}{2\pi r} = \frac{\frac{n\hbar}{m_e r}}{2\pi r} = \frac{n\hbar}{2\pi m_e r^2} = \frac{n\hbar}{2\pi m_e n^4 (a_0)^2} = \frac{\hbar}{2m_e(a_0)^2 n^3} = \left(\frac{m_e k^2 Z^2 e^4}{2\pi \hbar^3}\right)\left(\frac{1}{n^3}\right)$$

using $a_0 = \dfrac{\hbar^2}{m_e ke^2}$.

(b) $$\Delta E = h v_{\text{photon}} = \left(\frac{kZ^2 e^2}{2a_0 h}\right)\left(\frac{1}{n_f^2} - \frac{1}{n_i^2}\right)$$

$$f_{\text{photon}} = \left(\frac{kZ^2 e^2}{2a_0 h}\right)\left(\frac{n_i^2 - n_f^2}{n_f^2 n_i^2}\right) = \left(\frac{kZ^2 e^2}{2a_0 h}\right)\left[\frac{(n_i - n_f)(n_i - n_f)}{n_f^2 n_i^2}\right]$$

$$= \left(\frac{m_e K^2 Z^2 e^4}{2\pi \hbar^3}\right)\left(\frac{n_i + n_f}{2n_f^2 n_i^2}\right)(n_i - n_f)$$

For $n_i - n_f = 1$, $f_{\text{photon}} = \left(\dfrac{m_e k^2 Z^2 e^4}{2\pi \hbar^3}\right)\left(\dfrac{n_i + n_f}{2n_f^2 n_i^2}\right)$. For $n_i = 2$, $n_f = 1$, $n_i = 3$, $n_f = 2$, etc.,

$\dfrac{1}{n_i^3} < \dfrac{n_i + n_f}{2n_i^2 n_f^2} < \dfrac{1}{n_f^3}$.

(c)   Frequency of emitted radiation is *in between* the initial orbital frequency and the final. As $n_i \to \infty$ the initial and final orbital frequencies squeeze closer together making the frequency of emitted radiation equal to the orbital frequency. This result agrees with Bohr's correspondence principle.

4-33 $r_n = \dfrac{n^2 a_0}{Z} = \dfrac{n^2 \hbar^2}{mke^2 Z} = (1)^2 \dfrac{(1.05 \times 10^{-34} \text{ Js})^2}{(207 m_e)(8.99 \times 10^9 \text{ Nm}^2/\text{C})(1.60 \times 10^{-19} \text{ C})^2 (82)} = 3.1 \times 10^{-15}$ m. Note

that this means the muon grazes the nuclear surface, and so experiments with muonic atoms give information about the nuclear charge distribution.

$$E_n = -\frac{\left[(ke^2 Z^2)/(2a_0 n^2)\right]/ke^2 Z^2}{2\hbar^2 n^2/mke^2} = \frac{mk^2 e^4 Z^2}{2\hbar^2 n^2}$$

Using $m = 207 m_e$, $n = 1$, and $Z = 82$ yields $E_1 = 18.9$ MeV. (Using the reduced mass makes no difference in the answers to three significant figures.)

4-35 $\mu = \dfrac{m_e M}{m_e + M} = \dfrac{m_e}{2}$ since $m_e = M$. In general, $r_n = \dfrac{n^2 \hbar^2}{Z\mu ke^2}$, so for positronium $Z = 1$, $\mu = \dfrac{m_e}{2}$ and

$r_{\text{positronium}} = \left(\dfrac{\hbar^2 n^2}{1}\right)\left(\dfrac{m_e}{2}\right)ke^2 = 2a_0 n^2 = 2r_{\text{hydrogen}}$. Similarly, $E_{\text{positronium}} = \dfrac{E_{\text{hydrogen}}}{2} = \dfrac{-6.80 \text{ eV}}{n^2}$.

4-37 $hf = \Delta E = \dfrac{4\pi^2 m_e k^2 e^4}{2h^2}\left(\dfrac{1}{(n-1)^2} - \dfrac{1}{n^2}\right)$, $f = \left(\dfrac{2\pi^2 m_e k^2 e^4}{h^3}\right)\left(\dfrac{2n-1}{(n-1)^2(n^2)}\right)$ as $n \to \infty$,

$f \to \left(\dfrac{2\pi^2 m_e k^2 e^4}{h^3}\right)\left(\dfrac{2}{n^3}\right)$. The revolution frequency is $f = \dfrac{v}{2\pi r} = \left(\dfrac{1}{2\pi}\right)\left(\dfrac{ke^2}{m_e}\right)^{1/2} \cdot \left(\dfrac{1}{r^{3/2}}\right)$ where

$r = \dfrac{n^2 h^2}{4\pi^2 m_e ke^2}$ substituting for r, $f = \left(\dfrac{1}{2\pi}\right)\left(\dfrac{ke^2}{m_e}\right)^{1/2}\left(\dfrac{8\pi^3 m_e ke^3 (m_e k)^{1/2}}{n^3 h^3}\right) = \dfrac{4\pi^2 m_e k^2 e^4}{h^3 n^3}$.

4-39   (a)   Energy balance: $\dfrac{1}{2 m_e v^2} = \dfrac{1}{2}(m_e + M)V^2 + 4.9$. Conservation of momentum

$m_e v = (m_e + M)V$ where m_e = electron mass, v = electron speed, M = Hg mass, and

V = common speed of the electron and Hg after collision. Substituting $V = \dfrac{m_e v}{m_e + M}$

into the first equation gives: $\dfrac{1}{2 m_e v^2} = \dfrac{1}{2}(m_e + M)\dfrac{m_e^2 v^2}{(m_e + M)^2} + 4.9$ or $\dfrac{1}{2 m_e v^2} = \dfrac{4.9 \text{ eV}}{\frac{1-m_e}{m_e + M}}$.

As $\dfrac{m_e}{m_e + M} = 2.74 \times 10^{-6}$, $(m_e = 5.49 \times 10^{-4}$ u, $M = 200.6$ u$)$, $\left(\dfrac{1-m_e}{m_e + M}\right)^{-1} \approx \left(\dfrac{1+m_e}{m_e + M}\right)^{+1}$

and $\dfrac{1}{2 m_e v^2} \approx (4.9)\left(\dfrac{1+m_e}{m_e + M}\right)$eV $= (4.9)(1 + 2.74 \times 10^{-6}$ eV$) = 4.900\,013\,4$ eV.

(b) $v = \left(\dfrac{(2)(4.90 \text{ eV})}{m_e}\right)^{1/2} = \left(\dfrac{(2)(4.90 \text{ eV})}{0.511 \times 10^6 \text{ eV}/c^2}\right)^{1/2} = 4.38 \times 10^{-3} c = 1.31 \times 10^6$ m/s

(c) $V = \left(\dfrac{m_e}{m_e + M}\right)(v) = (2.74 \times 10^{-6})(1.31 \times 10^6 \text{ m/s}) = 3.60$ m/s

(d) $K_{e^-} = \dfrac{1}{2 m_e V^2} = (0.5)(0.511 \times 10^6 \text{ eV})\dfrac{(3.60 \text{ m/s})^2}{(3.00 \times 10^8 \text{ m/s})^2} = 3.68 \times 10^{-11}$ eV. Yes it is

justified to assume the electron loses all its kinetic energy.

4-41   Liquid O_2, $\lambda_{abs} = 1\,269$ nm, $E = \dfrac{hc}{\lambda} = \dfrac{1.239\,8 \times 10^{-6}}{1.269 \times 10^{-6}} = 0.977$ eV for each molecule. For two

molecules, $\lambda = \dfrac{hc}{2E} = 634$ nm, red. By absorbing the red photons, the liquid O_2 appears to be

blue.

4-43   (a)   Suppose the atoms move in the $+x$ direction. The absorption of a photon by an atom

is a completely inelastic collision, described by $mv_i\hat{\mathbf{i}} + \dfrac{h}{\lambda}(-\hat{\mathbf{i}}) = mv_f\hat{\mathbf{i}}$ so $v_f - v_i = -\dfrac{h}{m\lambda}$.

This happens promptly every time an atom has fallen back into the ground state, so it

happens every 10^{-8} s $= \Delta t$. Then,

$$a = \frac{v_f - v_i}{\Delta t} = -\frac{h}{m\lambda\Delta t} \sim -\frac{6.626 \times 10^{-34}\text{ J}\cdot\text{s}}{(10^{-25}\text{ kg})(500 \times 10^{-9}\text{ m})(10^{-8}\text{ s})} \sim -10^6\ \text{m}\big/\text{s}^2\,.$$

(b)   With constant average acceleration, $v_f^2 = v_i^2 + 2a\Delta x$, $0 \sim (10^3\ \text{m}/\text{s})^2 + 2(-10^6\ \text{m}/\text{s}^2)\Delta x$

so $\Delta x \sim \dfrac{(10^3\ \text{m}/\text{s})^2}{10^6\ \text{m}/\text{s}^2} \sim 1$ m.

5

Matter Waves

5-1 $\lambda = \dfrac{h}{p} = \dfrac{h}{mv} = \dfrac{6.63 \times 10^{-34} \text{ Js}}{1.67 \times 10^{-27} \text{ kg}} (10^6 \text{ m/s}) = 3.97 \times 10^{-13} \text{ m}$

5-3 $\lambda = \dfrac{h}{p} = \dfrac{h}{mv} = \dfrac{6.63 \times 10^{-34} \text{ Js}}{74 \text{ kg}} (5 \text{ m/s}) = 1.79 \times 10^{-36} \text{ m}$

5-5   (a) $\lambda = \dfrac{h}{p}$ or $p = \dfrac{h}{\lambda} = \dfrac{hc}{\lambda c} = \dfrac{1\,240 \text{ eV nm}}{(10 \text{ nm})(c)} = \dfrac{124 \text{ eV}}{c}$. As

$$K = E - mc^2 = \left[p^2 c^2 + \left(mc^2 \right)^2 \right]^{1/2} - mc^2,$$

we must use the relativistic expression for K, when $pc \approx mc^2$. Here

$pc = 124 \text{ eV} \ll mc^2 = 0.511 \text{ MeV}$, so we can use the classical expression for K, $K = \dfrac{p^2}{2m}$.

$$K = \dfrac{p^2}{2m} = \dfrac{p^2 c^2}{2mc^2} = \dfrac{(124 \text{ eV})^2}{2(0.511 \text{ MeV})} = 0.150 \text{ eV}$$

   (b)   Electrons with $\lambda = 0.10 \text{ nm}$ $p = \dfrac{hc}{\lambda c} = \dfrac{12\,400 \text{ eV}}{c}$ as in (a). As $pc \ll mc^2 = 0.511 \text{ MeV}$, use

$K = \dfrac{p^2}{2m} = p^2 c^2 = \dfrac{(12\,400)^2 (\text{eV})^2}{(2)(0.511 \times 10^6 \text{ eV})} = 150 \text{ eV}$.

   (c)   Electrons with $\lambda = 10 \text{ fm} = 10 \times 10^{-15} \text{ m}$, $p = \dfrac{hc}{\lambda c} = \dfrac{1.24 \times 10^3 \text{ MeV}}{c}$. As

$pc \gg mc^2 = 0.511 \text{ MeV}$, use

$$K = \left[p^2 c^2 + \left(mc^2 \right)^2 \right]^{1/2} - mc^2 = pc - mc^2 = 1\,240 \text{ MeV} - 0.511 \text{ MeV} = 1\,239 \text{ MeV}.$$

For alphas with $mc^2 = 3\,726$ MeV:

(a) p still is $\dfrac{124\text{ eV}}{c}$. As $pc \ll 3\,726$ MeV, we use $K = \dfrac{p^2}{2m}$:

$$K = \frac{p^2 c^2}{2mc^2} = \frac{(124\text{ eV})^2}{(2)(3\,726\text{ MeV})} = 2.06 \times 10^{-6}\text{ eV}.$$

(b) For alphas with $\lambda = 0.10$ nm, $p = \dfrac{12\,400\text{ eV}}{c}$. As $pc \ll mc^2 = 3\,726$ MeV,

$$K = \frac{p^2}{2m} = \frac{p^2 c^2}{2mc^2} = \frac{(12\,400\text{ eV})^2}{(2)(3\,726\text{ MeV})} = 0.020\,6\text{ eV}.$$

(c) $p = \dfrac{1.24 \times 10^3\text{ MeV}}{c}$ and $pc = 1\,240$ MeV $\sim mc^2 = 3\,726$ MeV. We use

$$K = \left[p^2 c^2 + \left(mc^2\right)^2\right]^{1/2} - mc^2 = \left[(1\,240\text{ MeV})^2 + (3\,726\text{ MeV})^2\right]^{1/2} - 3\,726\text{ MeV}$$
$$= 201\text{ MeV}.$$

5-7 A 10 MeV proton has $K = 10$ MeV $\ll 2mc^2 = 1\,877$ MeV so we can use the classical expression $p = (2mK)^{1/2}$. (See Problem 5-2)

$$\lambda = \frac{h}{p} = \frac{hc}{[(2)(938.3\text{ MeV})(10\text{ MeV})]^{1/2}} = \frac{1\,240\text{ MeV fm}}{\left[(2)(938.3)(10)(\text{MeV})^2\right]^{1/2}} = 9.05\text{ fm} = 9.05 \times 10^{-15}\text{ m}$$

5-9 $m = 0.20$ kg: $mgh = \dfrac{mv^2}{2}$: $v = (2gh)^{1/2}$

$$p = mv = m(2gh)^{1/2} = (0.20)[2(9.80)(50)]^{1/2} = 6.261\text{ kg} \cdot \text{m/s}$$

$$\lambda = \frac{h}{p} = \frac{6.626 \times 10^{-34}\text{ J} \cdot \text{s}}{6.261\text{ kg} \cdot \text{m/s}} = 1.06 \times 10^{-34}\text{ m}$$

5-11 (a) In this problem, the electron must be treated relativistically because we must use relativity when $pc \approx mc^2$. (See problem 5-5). the momentum of the electron is

$$p = \frac{h}{\lambda} = \frac{6.626 \times 10^{-34}\text{ J} \cdot \text{s}}{10^{-14}\text{ m}} = 6.626 \times 10^{-20}\text{ kg} \cdot \text{m/s}$$

and $pc = 124$ MeV $\gg mc^2 = 0.511$ MeV. The energy of the electron is

$$E = \left(p^2 c^2 + m^2 c^4\right)^{1/2}$$
$$= \left[\left(6.626 \times 10^{-20}\text{ kg} \cdot \text{m/s}\right)^2 \left(3 \times 10^8\text{ m/s}\right)^2 + \left(0.511 \times 10^6\text{ eV}\right)^2 \left(1.602 \times 10^{-19}\text{ J/eV}\right)^2\right]^{1/2}$$
$$= 1.99 \times 10^{-11}\text{ J} = 1.24 \times 10^8\text{ eV}$$

so that $K = E - mc^2 \approx 124$ MeV.

(b)   The kinetic energy is too large to expect that the electron could be confined to a region the size of the nucleus.

5-13   A canceling wave will be produced when the path length difference between the surface reflection and the reflection from the nth plane below the surface equals some whole number of wavelengths plus $\dfrac{\lambda}{2}$. As the path length difference between a surface reflection and a reflection from plane n is given by $(n)(1.01\lambda)$, we find that a reflection from the 50th plane has a path difference of 50.5λ with the surface reflection, and consequently cancels the surface reflection. Essentially all waves reflected at θ will cancel as the wave reflected from the second plane will be cancelled by a reflection from the 51st plane and so on.

5-15   For a free, non-relativistic electron $E = \dfrac{m_e v_0^2}{2} = \dfrac{p^2}{2m_e}$. As the wavenumber and angular frequency of the electron's de Broglie wave are given by $p = \hbar k$ and $E = \hbar\omega$, substituting these results gives the dispersion relation $\omega = \dfrac{\hbar k^2}{2m_e}$. So $v_g = \dfrac{d\omega}{dk} = \dfrac{\hbar k}{m_e} = \dfrac{p}{m_e} = v_0$.

5-17 $E^2 = p^2 c^2 + \left(m_e c^2\right)^2$

$E = \left[p^2 c^2 + \left(m_e c^2\right)^2\right]^{1/2}$. As $E = \hbar\omega$ and $p = \hbar k$

$\hbar\omega = \left[\hbar^2 k^2 c^2 + \left(m_e c^2\right)^2\right]^{1/2}$ or

$\omega(k) = \left[k^2 c^2 + \dfrac{\left(m_e c^2\right)^2}{\hbar^2}\right]^{1/2}$

$v_p = \dfrac{\omega}{k} = \dfrac{\left[k^2 c^2 + \left(m_e c^2/\hbar\right)^2\right]^{1/2}}{k} = \left[c^2 + \left(\dfrac{m_e c^2}{\hbar k}\right)^2\right]^{1/2}$

$v_g = \dfrac{d\omega}{dk}\bigg|_{k_0} = \dfrac{1}{2}\left[k^2 c^2 + \left(\dfrac{m_e c^2}{\hbar}\right)^2\right]^{-1/2} 2kc^2 = \dfrac{kc^2}{\left[k^2 c^2 + \left(m_e c^2/\hbar\right)^2\right]^{1/2}}$

$v_p v_g = \left\{\dfrac{\left[k^2 c^2 + \left(m_e c^2/\hbar\right)^2\right]^{1/2}}{k}\right\}\left\{\left[k^2 c^2 + \left(m_e c^2/\hbar\right)^2\right]^{1/2}\right\} = c^2$

Therefore, $v_g < c$ if $v_p > c$.

5-19 $K = \dfrac{mv^2}{2} = \dfrac{p^2}{2m} : \left(1\times 10^6 \text{ eV}\right)\left(1.6\times 10^{-19} \text{ J/eV}\right) = \dfrac{p^2}{2\left(1.67\times 10^{-27} \text{ kg}\right)} \Rightarrow p = 2.312\times 10^{-20} \text{ kg}\cdot\text{m/s}$,

$\Delta p = 0.05 p = 1.160\times 10^{-21} \text{ kg}\cdot\text{m/s}$ and $\Delta x \Delta p = \dfrac{h}{4\pi}$. Thus

$$\Delta x = \dfrac{6.63\times 10^{-34} \text{ J}\cdot\text{s}}{\left(1.16\times 10^{-21} \text{ kg}\cdot\text{m/s}\right)(4\pi)} = 4.56\times 10^{-14} \text{ m}.$$

Note that non-relativistic treatment has been used, which is justified because the kinetic energy is only $\dfrac{(1.6\times10^{-13})\times100\%}{1.50\times10^{-10}} = 0.11\%$ of the rest energy.

5-21   (a)   The woman tries to hold a pellet within some horizontal region Δx_i and directly above the spot on the floor. The uncertainty principle requires her to give a pellet some x velocity at least as large as $\Delta v_x = \dfrac{\hbar}{2m\Delta x_i}$. When the pellet hits the floor at time t, the total miss distance is $\Delta x_{\text{total}} = \Delta x_i + \Delta v_x t = \Delta x_i + \left(\dfrac{\hbar}{2m\Delta x_i}\right)\sqrt{\dfrac{2H}{g}}$. The minimum value of the function Δx_{total} occurs for $\dfrac{d(\Delta x_{\text{total}})}{d(\Delta x_i)} = 0$ or $1 - \dfrac{\hbar}{2m}\sqrt{\dfrac{2H}{g}}(\Delta x_i)^{-2} = 0$.

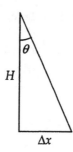

We find $\Delta x_i = \left(\dfrac{\hbar}{2m}\right)^{1/2}\left(\dfrac{2H}{g}\right)^{1/4}$.

(b)   For $H = 2.0$ m, $m = 0.50$ g, $\Delta x_{\text{total}} = 5.2\times10^{-16}$ m.

5-23   (a) $\Delta p \Delta x = m\Delta v \Delta x \geq \dfrac{\hbar}{2}$

$\Delta v \geq \dfrac{h}{4\pi m \Delta x} = \dfrac{2\pi J \cdot s}{4\pi(2\ \text{kg})(1\ \text{m})} = 0.25$ m/s

(b)   The duck might move by $(0.25\ \text{m/s})(5\ \text{s}) = 1.25$ m. With original position uncertainty of 1m, we can think of Δx growing to 1 m + 1.25 m = 2.25 m.

5-25   To find the energy width of the γ-ray use $\Delta E \Delta t \geq \dfrac{\hbar}{2}$ or

$$\Delta E \geq \dfrac{\hbar}{2\Delta t} \geq \dfrac{6.58\times10^{-16}\ \text{eV}\cdot\text{s}}{(2)(0.10\times10^{-9}\ \text{s})} \geq 3.29\times10^{-6}\ \text{eV}.$$

As the intrinsic energy width of $\sim \pm 3\times10^{-6}$ eV is so much less than the experimental resolution of ±5 eV, the intrinsic width can't be measured using this method.

5-27   For a single slit with width a, minima are given by $\sin\theta = \dfrac{n\lambda}{a}$ where $n = 1, 2, 3, \ldots$ and

$\sin\theta \approx \tan\theta = \dfrac{x}{L}$, $\dfrac{x_1}{L} = \dfrac{\lambda}{a}$ and $\dfrac{x_2}{L} = \dfrac{2\lambda}{a} \Rightarrow \dfrac{x_2 - x_1}{L} = \dfrac{\lambda}{a}$ or

$$\lambda = \frac{a\Delta x}{L} = \frac{5 \text{ Å} \times 2.1 \text{ cm}}{20 \text{ cm}} = 0.525 \text{ Å}$$

$$E = \frac{p^2}{2m} = \frac{h^2}{2m\lambda^2} = \frac{(hc)^2}{2mc^2\lambda^2} = \frac{\left(1.24 \times 10^4 \text{ eV} \cdot \text{Å}\right)^2}{2\left(5.11 \times 10^5 \text{ eV}\right)\left(0.525 \text{ Å}\right)^2} = 546 \text{ eV}$$

5-29   With *one* slit open $P_1 = |\Psi_1|^2$ or $P_2 = |\Psi_2|^2$. With <u>both</u> slits open, $P = |\Psi_1 + \Psi_2|^2$. At a maximum, the wavefunctions are in phase so

$$P_{\max} = \left(|\Psi_1| + |\Psi_2|\right)^2.$$

At a minimum, the wavefunctions are out of phase and

$$P_{\min} = \left(|\Psi_1| - |\Psi_2|\right)^2.$$

Now $\dfrac{P_1}{P_2} = \dfrac{|\Psi_1|^2}{|\Psi_2|^2} = 25$ or $\dfrac{|\Psi_1|}{|\Psi_2|} = 5$, and

$$\frac{P_{\max}}{P_{\min}} = \frac{\left(|\Psi_1| + |\Psi_2|\right)^2}{\left(|\Psi_1| - |\Psi_2|\right)^2} = \frac{\left(5|\Psi_2| + |\Psi_2|\right)^2}{\left(5|\Psi_1| - |\Psi_2|\right)^2} = \frac{6^2}{4^2} = \frac{36}{16} = 2.25.$$

5-31

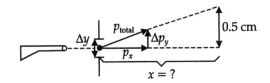

$\Delta y \Delta p_y \sim \hbar$ $\Delta p_y = \dfrac{\hbar}{\Delta y}$. From the diagram, because the momentum triangle and space triangle

are similar, $\dfrac{\Delta p_y}{p_x} = \dfrac{0.5 \text{ cm}}{x}$;

$$x = \frac{(0.5 \text{ cm})p_x}{\Delta p_y} = \frac{(0.5 \text{ cm})p_x \Delta y}{\hbar} = \frac{\left(0.5 \times 10^{-2} \text{ m}\right)\left(0.001 \text{ kg}\right)\left(100 \text{ m/s}\right)\left(2 \times 10^{-3} \text{ m}\right)}{1.05 \times 10^{-34} \text{ J} \cdot \text{s}}$$

$$= 9.5 \times 10^{27} \text{ m}$$

Once again we see that the uncertainty relation has no observable consequences for macroscopic systems.

5-33 From the uncertainty principle, $\Delta E\Delta t \sim \hbar$ $\Delta mc^2\Delta t = \hbar$. Therefore,

$$\frac{\Delta m}{m} = \frac{h}{2\pi c^2\Delta tm} = \frac{h}{2\pi\Delta tE_{rest}} = \frac{6.63\times10^{-34}\ \text{J}\cdot\text{s}}{2\pi\left(8.7\times10^{-17}\ \text{s}\right)\left(135\times10^6\ \text{eV}\right)\left(1.6\times10^{-19}\ \text{J/eV}\right)} = 5.62\times10^{-8}.$$

5-35   (a) $f(x) = \dfrac{1}{\sqrt{2\pi}}\displaystyle\int_{-\infty}^{+\infty} a(k)e^{ikx}\,dk = \dfrac{A}{\sqrt{2\pi}}\int_{-\infty}^{+\infty} e^{-\alpha^2(k-k_0)^2}e^{ikx}\,dk = \dfrac{A}{\sqrt{2\pi}}e^{-\alpha^2k_0^2}\int_{-\infty}^{+\infty} e^{-\alpha^2\left(k^2-\left(2k_0+ix/\alpha^2\right)k\right)}\,dk.$

Now complete the square in order to get the integral into the standard form

$$\int_{-\infty}^{+\infty} e^{-az^2}\,dz:$$

$$e^{-\alpha^2\left(k^2-\left(2k_0+ix/\alpha^2\right)k\right)} = e^{+\alpha^2\left(k_0+ix/2\alpha^2\right)^2}e^{-\alpha^2\left(k-\left(k_0+ix/2\alpha^2\right)\right)^2}$$

$$f(x) = \frac{A}{\sqrt{2\pi}}e^{-\alpha^2k_0^2}e^{\alpha^2\left(k_0+ix/2\alpha^2\right)^2}\int_{k=-\infty}^{+\infty} e^{-\alpha^2\left(k-\left(k_0+ix/2\alpha^2\right)\right)^2}\,dk$$

$$= \frac{A}{\sqrt{2\pi}}e^{-x^2/4\alpha^2}e^{ik_0x}\int_{z=-\infty}^{+\infty} e^{-\alpha^2z^2}\,dz$$

where $z = k-\left(k_0+\dfrac{ix}{2\alpha^2}\right)$. Since $\displaystyle\int_{z=-\infty}^{+\infty} e^{-\alpha^2z^2}\,dz = \frac{\pi^{1/2}}{\alpha}$, $f(x) = \dfrac{A}{\alpha\sqrt{2}}e^{-x^2/4\alpha^2}e^{ik_0x}$. The real

part of $f(x)$, Re $f(x)$ is Re $f(x) = \dfrac{A}{\alpha\sqrt{2}}e^{-x^2 4\alpha^2}\cos k_0x$ and is a gaussian envelope

multiplying a harmonic wave with wave number k_0. A plot of Re $f(x)$ is shown
below:

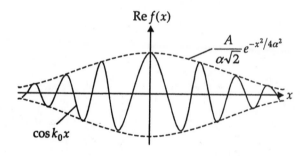

Comparing $\dfrac{A}{\alpha\sqrt{2}}e^{-x^2 4\alpha^2}$ to $Ae^{-(x/2\Delta x)^2}$ implies $\Delta x = \alpha$.

(c)   By same reasoning because $\alpha^2 = \dfrac{1}{4\Delta k^2}$, $\Delta k = \dfrac{1}{2\alpha}$. Finally $\Delta x\Delta k = \alpha\left(\dfrac{1}{2\alpha}\right) = \dfrac{1}{2}$.

5-37   We find the speed of each electron from energy conservation in the firing process:

$$0 = K_f + U_f = \frac{1}{2}mv^2 - eV$$

$$v = \sqrt{\frac{2eV}{m}} = \sqrt{\frac{2\left(1.6\times10^{-19}\ \text{C}\right)\left(45\ \text{V}\right)}{9.11\times10^{-31}\ \text{kg}}} = 3.98\times10^6\ \text{m/s}$$

The time of flight is $\Delta t = \dfrac{\Delta x}{v} = \dfrac{0.28 \text{ m}}{3.98 \times 10^6 \text{ m/s}} = 7.04 \times 10^{-8}$ s . The current when electrons are

28 cm apart is $I = \dfrac{q}{t} = \dfrac{e}{\Delta t} = \dfrac{1.6 \times 10^{-19} \text{ C}}{7.04 \times 10^{-8} \text{ s}} = 2.27 \times 10^{-12}$ A .

6

Quantum Mechanics in One Dimension

6-1   (a)   Not acceptable – diverges as $x \to \infty$.

       (b)   Acceptable.

       (c)   Acceptable.

       (d)   Not acceptable – not a single-valued function.

       (e)   Not acceptable – the wave is discontinuous (as is the slope).

6-3   (a) $A\sin\left(\dfrac{2\pi x}{\lambda}\right) = A\sin\left(5\times10^{10}\,x\right)$ so $\left(\dfrac{2\pi}{\lambda}\right) = 5\times10^{10}$ m^{-1}, $\lambda = \dfrac{2\pi}{5\times10^{10}} = 1.26\times10^{-10}$ m.

       (b) $p = \dfrac{h}{\lambda} = \dfrac{6.626\times10^{-34}\ \text{Js}}{1.26\times10^{-10}\ \text{m}} = 5.26\times10^{-24}$ kg m/s

       (c) $K = \dfrac{p^2}{2m}$ $m = 9.11\times10^{-31}$ kg

               $K = \dfrac{\left(5.26\times10^{-24}\ \text{kg m/s}\right)^2}{\left(2\times9.11\times10^{-31}\ \text{kg}\right)} = 1.52\times10^{-17}$ J

               $K = \dfrac{1.52\times10^{-17}\ \text{J}}{1.6\times10^{-19}\ \text{J/eV}} = 95$ eV

6-5   (a)   Solving the Schrödinger equation for U with $E = 0$ gives

$$U = \left(\frac{\hbar^2}{2m}\right)\frac{\left(\frac{d^2\psi}{dx^2}\right)}{\psi}.$$

           If $\psi = Ae^{-x^2/L^2}$ then $\dfrac{d^2\psi}{dx^2} = \left(4Ax^3 - 6AxL^2\right)\left(\dfrac{1}{L^4}\right)e^{-x^2/L^2}$, $U = \left(\dfrac{\hbar^2}{2mL^2}\right)\left(\dfrac{4x^2}{L^2} - 6\right)$.

(b) $U(x)$ is a parabola centered at $x = 0$ with $U(0) = \dfrac{-3\hbar^2}{mL^2} < 0$:

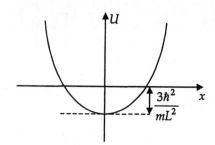

6-7   Since the particle is confined to the box, Δx can be no larger than L, the box length. With $n = 0$, the particle energy $E_n = \dfrac{n^2 h^2}{8mL^2}$ is also zero. Since the energy is all kinetic, this implies $\langle p_x^2 \rangle = 0$. But $\langle p_x \rangle = 0$ is expected for a particle that spends equal time moving left as right, giving $\Delta p_x = \sqrt{\langle p_x^2 \rangle - \langle p_x \rangle^2} = 0$. Thus, for this case $\Delta p_x \Delta x = 0$, in violation of the uncertainty principle.

6-9 $E_n = \dfrac{n^2 h^2}{8mL^2}$, so $\Delta E = E_2 - E_1 = \dfrac{3h^2}{8mL^2}$

$$\Delta E = (3)\frac{(1240 \text{ eV nm}/c)^2}{8(938.28 \times 10^6 \text{ eV}/c^2)(10^{-5} \text{ nm})^2} = 6.14 \text{ MeV}$$

$$\lambda = \frac{hc}{\Delta E} = \frac{1240 \text{ eV nm}}{6.14 \times 10^6 \text{ eV}} = 2.02 \times 10^{-4} \text{ nm}$$

This is the gamma ray region of the electromagnetic spectrum.

6-11   In the present case, the box is displaced from $(0, L)$ by $\dfrac{L}{2}$. Accordingly, we may obtain the wavefunctions by replacing x with $x - \dfrac{L}{2}$ in the wavefunctions of Equation 6.18. Using

$$\sin\left[\left(\frac{n\pi}{L}\right)\left(x - \frac{L}{2}\right)\right] = \sin\left[\left(\frac{n\pi x}{L}\right) - \frac{n\pi}{2}\right] = \sin\left(\frac{n\pi x}{L}\right)\cos\left(\frac{n\pi}{2}\right) - \cos\left(\frac{n\pi x}{L}\right)\sin\left(\frac{n\pi}{2}\right)$$

we get for $-\dfrac{L}{2} \le x \le \dfrac{L}{2}$

$$\psi_1(x) = \left(\frac{2}{L}\right)^{1/2} \cos\left(\frac{\pi x}{L}\right); \quad P_1(x) = \left(\frac{2}{L}\right)\cos^2\left(\frac{\pi x}{L}\right)$$

$$\psi_2(x) = \left(\frac{2}{L}\right)^{1/2} \sin\left(\frac{2\pi x}{L}\right); \quad P_2(x) = \left(\frac{2}{L}\right)\sin^2\left(\frac{2\pi x}{L}\right)$$

$$\psi_3(x) = \left(\frac{2}{L}\right)^{1/2} \cos\left(\frac{3\pi x}{L}\right); \quad P_3(x) = \left(\frac{2}{L}\right)\cos^2\left(\frac{3\pi x}{L}\right)$$

6-13 **(a)** Proton in a box of width $L = 0.200$ nm $= 2 \times 10^{-10}$ m

$$E_1 = \frac{h^2}{8m_pL^2} = \frac{\left(6.626 \times 10^{-34} \text{ J} \cdot \text{s}\right)^2}{8\left(1.67 \times 10^{-27} \text{ kg}\right)\left(2 \times 10^{-10} \text{ m}\right)^2} = 8.22 \times 10^{-22} \text{ J}$$

$$= \frac{8.22 \times 10^{-22} \text{ J}}{1.60 \times 10^{-19} \text{ J/eV}} = 5.13 \times 10^{-3} \text{ eV}$$

(b) Electron in the same box:

$$E_1 = \frac{h^2}{8m_eL^2} = \frac{\left(6.626 \times 10^{-34} \text{ J} \cdot \text{s}\right)^2}{8\left(9.11 \times 10^{-31} \text{ kg}\right)\left(2 \times 10^{-10} \text{ m}\right)^2} = 1.506 \times 10^{-18} \text{ J} = 9.40 \text{ eV}.$$

(c) The electron has a much higher energy because it is much less massive.

6-15 **(a)** $U = \left(\dfrac{e^2}{4\pi\varepsilon_0 d}\right)\left[-1 + \dfrac{1}{2} - \dfrac{1}{3} + \left(-1 + \dfrac{1}{2}\right) + (-1)\right] = \dfrac{(-7/3)e^2}{4\pi\varepsilon_0 d} = \dfrac{(-7/3)ke^2}{d}$

(b) $K = 2E_1 = \dfrac{2h^2}{8m \times 9d^2} = \dfrac{h^2}{36md^2}$

(c) $E = U + K$ and $\dfrac{dE}{dd} = 0$ for a minimum $\left[\dfrac{(+7/3)e^2k}{d^2}\right] - \dfrac{h^2}{18md^3} = 0$

$d = \dfrac{3h^2}{(7)\left(18ke^2m\right)}$ or $d = \dfrac{h^2}{42mke^2}$

$d = \dfrac{\left(6.63 \times 10^{-34} \text{ J} \cdot \text{s}\right)^2}{(42)\left(9.11 \times 10^{-31} \text{ kg}\right)\left(9 \times 10^9 \text{ N} \cdot \text{m}^2 \cdot \text{C}^{-2}\right)\left(1.6 \times 10^{-19} \text{ C}\right)^2} = 0.5 \times 10^{-10} \text{ m} = 0.050 \text{ nm}$

(d) Since the lithium spacing is a, where $Na^3 = V$ and the density is $\dfrac{Nm}{V}$ where m is the mass of one atom, we get

$$a = \left(\frac{Vm}{Nm}\right)^{1/3} = \left(\frac{m}{\text{density}}\right)^{1/3} = \left(1.66 \times 10^{-27} \text{ kg} \times \frac{7}{530 \text{ kg/m}^3}\right)^{1/3} \quad m = 2.8 \times 10^{-10} \text{ m}$$

$$= 0.28 \text{ nm}$$

(2.8 times larger than $2d$)

6-17 **(a)** The wavefunctions and probability densities are the same as those shown in the two lower curves in Figure 6.16 of the text.

(b) $P_1 = \displaystyle\int_{1.5\text{ Å}}^{3.5\text{ Å}} |\psi|^2 \, dx = \frac{2}{10 \text{ Å}} \int_{1.5\text{ Å}}^{3.5\text{ Å}} \sin^2\left(\frac{\pi x}{10}\right) dx$

$$\frac{1}{5}\left[\frac{x}{2} - \frac{10}{4\pi}\sin\left(\frac{\pi x}{5}\right)\right]_{1.5}^{3.5}$$

In the above result we used $\int \sin^2 ax\,dx = \dfrac{x}{2} - \dfrac{1}{4a}\sin(2ax)$. Therefore,

$$P_1 = \frac{1}{10}\left[x - \frac{5}{\pi}\sin\left(\frac{\pi x}{5}\right)\right]_{1.5}^{3.5} = \frac{1}{10}\left\{3.5 - \frac{5}{\pi}\sin\left[\frac{\pi(3.5)}{5}\right] - 1.5 + \frac{5}{\pi}\sin\left[\frac{\pi(1.5)}{5}\right]\right\}$$

$$= \frac{1}{10}\left[2.0 + \frac{5}{\pi}(\sin 0.3\pi - \sin 0.7\pi)\right] = \frac{1}{10}[2.00 + 0.0] = 0.200$$

(c) $$P_2 = \frac{1}{5}\int_{1.5}^{3.5}\sin^2\left(\frac{\pi x}{5}\right)dx = \frac{1}{5}\left[\frac{x}{2} - \frac{5}{4\pi}\sin(0.4\pi x)\right]_{1.5}^{3.5} = \frac{1}{10}\left[x - \frac{5}{2\pi}\sin(0.4\pi x)\right]_{1.5}^{3.5}$$

$$= \frac{1}{10}\{2.0 + (0.798)\{\sin[0.4\pi(1.5)] - \sin[0.4\pi(3.5)]\}\} = 0.351$$

(d)   Using $E = \dfrac{n^2 h^2}{8mL^2}$ we find $E_1 = 0.377$ eV and $E_2 = 1.51$ eV.

6-19   The allowed energies for this system are given by Equation 6.17, or $E_n = \dfrac{n^2\pi^2\hbar^2}{2mL^2} = \dfrac{n^2 h^2}{8mL^2}$.

Using $E_n = 10^{-3}$ J, $m = 10^{-3}$ kg, $L = 10^{-2}$ m and solving for n gives

$$n = \frac{\left\{8(10^{-3}\text{ kg})(10^{-2}\text{ m})^2(10^{-3}\text{ J})\right\}^{1/2}}{6.63\times 10^{-34}\text{ J}\cdot\text{s}} = 4.27\times 10^{28}.$$

The excitation energy is $\Delta E = E_{n+1} - E_n$, or

$$\Delta E = \frac{h^2}{8mL^2}\left\{(n+1)^2 - n^2\right\} = \left(\frac{h^2}{8mL^2}\right)\{2n+1\} = E_n\left(\frac{2n+1}{n^2}\right) \approx \frac{2}{n}E_n \text{ for } n \gg 1.$$

Thus, $\Delta E \approx \dfrac{(2)(10^{-3}\text{ J})}{4.27\times 10^{28}} = 4.69\times 10^{-32}$ J.

6-21 $n = 4$

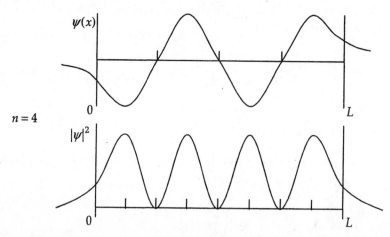

$n = 4$

Note that the $n = 4$ wavefunction has three nodes and is antisymmetric about the midpoint of the well.

6-23   Inside the well, the particle is free and the Schrödinger waveform is trigonometric with wavenumber $k = \left(\dfrac{2mE}{\hbar^2}\right)^{1/2}$:

$$\psi(x) = A\sin kx + B\cos kx \quad 0 \le x \le L.$$

The infinite wall at $x = 0$ requires $\psi(0) = B = 0$. Beyond $x = L$, $U(x) = U$ and the Schrödinger equation $\dfrac{d^2\psi}{dx^2} = \left(\dfrac{2m}{\hbar^2}\right)\{U - E\}\psi(x)$, which has exponential solutions for $E < U$

$$\psi(x) = Ce^{-\alpha x} + De^{+\alpha x}, \qquad x > L$$

where $\alpha = \left[\dfrac{2m(U - E)}{\hbar^2}\right]^{1/2}$. To keep ψ bounded at $x = \infty$ we must take $D = 0$. At $x = L$, continuity of ψ and $\dfrac{d\psi}{dx}$ demands

$$A\sin kL = Ce^{-\alpha L}$$
$$kA\cos kL = -\alpha Ce^{-\alpha L}$$

Dividing one by the other gives an equation for the allowed particle energies: $k\cot kL = -\alpha$. The dependence on E (or k) is made more explicit by noting that $k^2 + \alpha^2 = \dfrac{2mU}{\hbar^2}$, which allows the energy condition to be written $k\cot kL = -\left[\left(\dfrac{2mU}{\hbar^2}\right) - k^2\right]^{1/2}$. Multiplying by L, squaring the result, and using $\cot^2\theta + 1 = \csc^2\theta$ gives $(kL)^2\csc^2(kL) = \dfrac{2mUL^2}{\hbar^2}$ from which we obtain $\dfrac{kL}{\sin kL} = \left(\dfrac{2mUL^2}{\hbar^2}\right)^{1/2}$. Since $\dfrac{\theta}{\sin\theta}$ is never smaller than unity for *any* value of θ, there can be no bound state energies if $\dfrac{2mUL^2}{\hbar^2} < 1$.

6-25   At its limits of vibration $x = \pm A$ the classical oscillator has all its energy in potential form: $E = \dfrac{1}{2}m\omega^2 A^2$ or $A = \left(\dfrac{2E}{m\omega^2}\right)^{1/2}$. If the energy is quantized as $E_n = \left(n + \dfrac{1}{2}\right)\hbar\omega$, then the corresponding amplitudes are $A_n = \left[\dfrac{(2n + 1)\hbar}{m\omega}\right]^{1/2}$.

6-29   (a)   Normalization requires $1 = \int\limits_{-\infty}^{\infty}|\psi|^2\,dx = C^2\int\limits_{0}^{\infty}e^{-2x}\left(1 - e^{-x}\right)^2 dx = C^2\int\limits_{0}^{\infty}\left(e^{-2x} - 2e^{-3x} + e^{-4x}\right)dx.$

The integrals are elementary and give $1 = C^2\left\{\dfrac{1}{2} - 2\left(\dfrac{1}{3}\right) + \dfrac{1}{4}\right\} = \dfrac{C^2}{12}$. The proper units for C are those of $(\text{length})^{-1/2}$ thus, normalization requires $C = (12)^{1/2}\ \text{nm}^{-1/2}$.

(b)   The most likely place for the electron is where the probability $|\psi|^2$ is largest. This is also where ψ itself is largest, and is found by setting the derivative $\dfrac{d\psi}{dx}$ equal zero:

$$0 = \frac{d\psi}{dx} = C\{-e^{-x} + 2e^{-2x}\} = Ce^{-x}\{2e^{-x} - 1\}.$$

The RHS vanishes when $x = \infty$ (a minimum), and when $2e^{-x} = 1$, or $x = \ln 2$ nm. Thus, the most likely position is at $x_p = \ln 2$ nm $= 0.693$ nm.

(c)   The average position is calculated from

$$\langle x \rangle = \int\limits_{-\infty}^{\infty} x|\psi|^2\, dx = C^2 \int\limits_{0}^{\infty} xe^{-2x}\left(1 - e^{-x}\right)^2 dx = C^2 \int\limits_{0}^{\infty} x\left(e^{-2x} - 2e^{-3x} + e^{-4x}\right) dx.$$

The integrals are readily evaluated with the help of the formula $\int\limits_{0}^{\infty} xe^{-ax}\, dx = \dfrac{1}{a^2}$ to get

$\langle x \rangle = C^2\left\{\dfrac{1}{4} - 2\left(\dfrac{1}{9}\right) + \dfrac{1}{16}\right\} = C^2\left\{\dfrac{13}{144}\right\}$. Substituting $C^2 = 12$ nm^{-1} gives

$$\langle x \rangle = \frac{13}{12}\ \text{nm} = 1.083\ \text{nm}.$$

We see that $\langle x \rangle$ is somewhat greater than the most probable position, since the probability density is skewed in such a way that values of x larger than x_p are weighted more heavily in the calculation of the average.

6-31   The symmetry of $|\psi(x)|^2$ about $x = 0$ can be exploited effectively in the calculation of average values. To find $\langle x \rangle$

$$\langle x \rangle = \int\limits_{-\infty}^{\infty} x|\psi(x)|^2\, dx$$

We notice that the integrand is antisymmetric about $x = 0$ due to the extra factor of x (an odd function). Thus, the contribution from the two half-axes $x > 0$ and $x < 0$ cancel exactly, leaving $\langle x \rangle = 0$. For the calculation of $\langle x^2 \rangle$, however, the integrand is symmetric and the half-axes contribute equally to the value of the integral, giving

$$\langle x \rangle = \int\limits_{0}^{\infty} x^2|\psi|^2\, dx = 2C^2 \int\limits_{0}^{\infty} x^2 e^{-2x/x_0}\, dx.$$

Two integrations by parts show the value of the integral to be $2\left(\dfrac{x_0}{2}\right)^3$. Upon substituting for

C^2, we get $\langle x^2 \rangle = 2\left(\dfrac{1}{x_0}\right)(2)\left(\dfrac{x_0}{2}\right)^3 = \dfrac{x_0^2}{2}$ and $\Delta x = \left(\langle x^2 \rangle - \langle x \rangle^2\right)^{1/2} = \left(\dfrac{x_0^2}{2}\right)^{1/2} = \dfrac{x_0}{\sqrt{2}}$. In calculating

the probability for the interval $-\Delta x$ to $+\Delta x$ we appeal to symmetry once again to write

$$P = \int_{-\Delta x}^{+\Delta x} |\psi|^2 \, dx = 2C^2 \int_0^{\Delta x} e^{-2x/x_0} \, dx = -2C^2 \left(\dfrac{x_0}{2}\right) e^{-2x/x_0}\Big|_0^{\Delta x} = 1 - e^{-\sqrt{2}} = 0.757$$

or about 75.7% independent of x_0.

6-33   (a)   Since there is no preference for motion in the leftward sense vs. the rightward sense, a particle would spend equal time moving left as moving right, suggesting $\langle p_x \rangle = 0$.

(b)   To find $\langle p_x^2 \rangle$ we express the average energy as the sum of its kinetic and potential energy contributions: $\langle E \rangle = \left\langle \dfrac{p_x^2}{2m}\right\rangle + \langle U \rangle = \dfrac{\langle p_x^2 \rangle}{2m} + \langle U \rangle$. But energy is sharp in the oscillator ground state, so that $\langle E \rangle = E_0 = \dfrac{1}{2}\hbar\omega$. Furthermore, remembering that $U(x) = \dfrac{1}{2}m\omega^2 x^2$ for the quantum oscillator, and using $\langle x^2 \rangle = \dfrac{\hbar}{2m\omega}$ from Problem 6-32, gives $\langle U \rangle = \dfrac{1}{2}m\omega^2\langle x^2 \rangle = \dfrac{1}{4}\hbar\omega$. Then $\langle p_x^2 \rangle = 2m(E_0 - \langle U \rangle) = 2m\left(\dfrac{\hbar\omega}{4}\right) = \dfrac{m\hbar\omega}{2}$.

(c) $\Delta p_x = \left(\langle p_x^2 \rangle - \langle p_x \rangle^2\right)^{1/2} = \left(\dfrac{m\hbar\omega}{2}\right)^{1/2}$

6-35   Applying the momentum operator $[p_x] = \left(\dfrac{\hbar}{i}\right)\dfrac{d}{dx}$ to each of the candidate functions yields

(a) $[p_x]\{A\sin(kx)\} = \left(\dfrac{\hbar}{i}\right)k\{A\cos(kx)\}$

(b) $[p_x]\{A\sin(kx) - A\cos(kx)\} = \left(\dfrac{\hbar}{i}\right)k\{A\cos(kx) + A\sin(kx)\}$

(c) $[p_x]\{A\cos(kx) + iA\sin(kx)\} = \left(\dfrac{\hbar}{i}\right)k\{-A\sin(kx) + iA\cos(kx)\}$

(d) $[p_x]\{e^{ik(x-a)}\} = \left(\dfrac{\hbar}{i}\right)ik\{e^{ik(x-a)}\}$

In case (c), the result is a multiple of the original function, since

$$-A\sin(kx) + iA\cos(kx) = i\{A\cos(kx) + iA\sin(kx)\}.$$

The multiple is $\left(\dfrac{\hbar}{i}\right)(ik) = \hbar k$ and is the eigenvalue. Likewise for (d), the operation $[p_x]$ returns the original function with the multiplier $\hbar k$. Thus, (c) and (d) are eigenfunctions of $[p_x]$ with eigenvalue $\hbar k$, whereas (a) and (b) are not eigenfunctions of this operator.

6-37   (a)   Normalization requires

$$1 = \int_{-\infty}^{\infty} |\Psi|^2 \, dx = C^2 \int_{-\infty}^{\infty} \{\psi_1^* + \psi_2^*\}\{\psi_1 + \psi_2\} \, dx$$

$$= C^2 \left\{ \int |\psi_1|^2 \, dx + \int |\psi_2|^2 \, dx + \int \psi_2^* \psi_1 \, dx + \int \psi_1^* \psi_2 \, dx \right\}.$$

The first two integrals on the right are unity, while the last two are, in fact, the same integral since ψ_1 and ψ_2 are both real. Using the waveforms for the infinite square well, we find

$$\int \psi_2 \psi_1 \, dx = \frac{2}{L} \int_0^L \sin\left(\frac{\pi x}{L}\right) \sin\left(\frac{2\pi x}{L}\right) dx = \frac{1}{L} \int_0^L \left\{ \cos\left(\frac{\pi x}{L}\right) - \cos\left(\frac{3\pi x}{L}\right) \right\} dx$$

where, in writing the last line, we have used the trigonometric exponential identities of sine and cosine. Both of the integrals remaining are readily evaluated, and are zero. Thus, $1 = C^2\{1 + 0 + 0 + 0\} = 2C^2$, or $C = \dfrac{1}{\sqrt{2}}$. Since $\psi_{1,2}$ are stationary states, they develop in time according to their respective energies $E_{1,2}$ as $e^{-iEt/\hbar}$. Then $\Psi(x, t) = C\{\psi_1 e^{-iE_1 t/\hbar} + \psi_2 e^{-iE_2 t/\hbar}\}$.

(c) $\Psi(x, t)$ is a stationary state only if it is an eigenfunction of the energy operator $[E] = i\hbar \dfrac{\partial}{\partial t}$. Applying $[E]$ to Ψ gives

$$[E]\Psi = C\left\{ i\hbar\left(\frac{-iE_1}{\hbar}\right)\psi_1 e^{-iE_1 t/\hbar} + i\hbar\left(\frac{-iE_2}{\hbar}\right)\psi_2 e^{-iE_2 t/\hbar} \right\} = C\{E_1 \psi_1 e^{-iE_1 t/\hbar} + E_2 \psi_2 e^{-iE_2 t/\hbar}\}.$$

Since $E_1 \neq E_2$, the operations $[E]$ does *not* return a multiple of the wavefunction, and so Ψ is not a stationary state. Nonetheless, we may calculate the average energy for this state as

$$\langle E \rangle = \int \Psi^* [E]\Psi \, dx = C^2 \int \{\psi_1^* e^{+iE_1 t/\hbar} + \psi_2^* e^{+iE_2 t/\hbar}\}\{E_1 \psi_1 e^{-iE_1 t/\hbar} + E_2 \psi_2 e^{-iE_2 t/\hbar}\} \, dx$$

$$= C^2 \left\{ E_1 \int |\psi_1|^2 \, dx + E_2 \int |\psi_2|^2 \, dx \right\}$$

with the cross terms vanishing as in part (a). Since $\psi_{1,2}$ are normalized and $C^2 = \dfrac{1}{2}$ we get finally $\langle E \rangle = \dfrac{E_1 + E_2}{2}$.

7

Tunneling Phenomena

7-1 (a) The reflection coefficient is the ratio of the reflected intensity to the incident wave intensity, or $R = \dfrac{|(1/2)(1-i)|^2}{|(1/2)(1+i)|^2}$. But $|1-i|^2 = (1-i)(1-i)^* = (1-i)(1+i) = |1+i|^2 = 2$, so that $R = 1$ in this case.

(b) To the left of the step the particle is free. The solutions to Schrödinger's equation are $e^{\pm ikx}$ with wavenumber $k = \left(\dfrac{2mE}{\hbar^2}\right)^{1/2}$. To the right of the step $U(x) = U$ and the equation is $\dfrac{d^2\psi}{dx^2} = \dfrac{2m}{\hbar^2}(U-E)\psi(x)$. With $\psi(x) = e^{-kx}$, we find $\dfrac{d^2\psi}{dx^2} = k^2\psi(x)$, so that $k = \left[\dfrac{2m(U-E)}{\hbar^2}\right]^{1/2}$. Substituting $k = \left(\dfrac{2mE}{\hbar^2}\right)^{1/2}$ shows that $\left[\dfrac{E}{(U-E)}\right]^{1/2} = 1$ or $\dfrac{E}{U} = \dfrac{1}{2}$.

(c) For 10 MeV protons, $E = 10$ MeV and $m = \dfrac{938.28 \text{ MeV}}{c^2}$. Using $\hbar = 197.3$ MeV fm/c $(1 \text{ fm} = 10^{-15} \text{ m})$, we find
$$\delta = \frac{1}{k} = \frac{\hbar}{(2mE)^{1/2}} = \frac{197.3 \text{ MeV fm}/c}{\left[(2)(938.28 \text{ MeV}/c^2)(10 \text{ MeV})\right]^{1/2}} = 1.44 \text{ fm}.$$

7-3 With $E = 25$ MeV and $U = 20$ MeV, the ratio of wavenumber is
$\dfrac{k_1}{k_2} = \left(\dfrac{E}{E-U}\right)^{1/2} = \left(\dfrac{25}{25-20}\right)^{1/2} = \sqrt{5} = 2.236$. Then from Problem 7-2 $R = \dfrac{(\sqrt{5}-1)^2}{(\sqrt{5}+1)^2} = 0.146$ and

$T = 1 - R = 0.854$. Thus, 14.6% of the incoming particles would be reflected and 85.4% would be transmitted. For electrons with the same energy, the transparency and reflectivity of the step are unchanged.

7-5 (a) The transmission probability according to Equation 7.9 is
$\dfrac{1}{T(E)} = 1 + \left[\dfrac{U^2}{4E(U-E)}\right]\sinh^2 \alpha L$ with $\alpha = \dfrac{[2m(U-E)]^{1/2}}{\hbar}$. For $E \ll U$, we find $(\alpha L)^2 \approx \dfrac{2mUL^2}{\hbar^2} \gg 1$ by hypothesis. Thus, we may write $\sinh \alpha L \approx \dfrac{1}{2}e^{\alpha L}$. Also $U - E \approx U$, giving $\dfrac{1}{T(E)} \approx 1 + \left(\dfrac{U}{16E}\right)e^{2\alpha L} \approx \left(\dfrac{U}{16E}\right)e^{2\alpha L}$ and a probability for transmission $P = T(E) = \left(\dfrac{16E}{U}\right)e^{-2\alpha L}$.

(b)   Numerical Estimates: $(\hbar = 1.055 \times 10^{-34}$ Js$)$

1)   For $m = 9.11 \times 10^{-31}$ kg, $U - E = 1.60 \times 10^{-21}$ J, $L = 10^{-10}$ m;
$$\alpha = \frac{[2m(U-E)]^{1/2}}{\hbar} = 5.12 \times 10^8 \text{ m}^{-1} \text{ and } e^{-2\alpha L} = 0.90$$

2)   For $m = 9.11 \times 10^{-31}$ kg, $U - E = 1.60 \times 10^{-19}$ J, $L = 10^{-10}$ m; $\alpha = 5.12 \times 10^9$ m^{-1} and $e^{-2\alpha L} = 0.36$

3)   For $m = 6.7 \times 10^{-27}$ kg, $U - E = 1.60 \times 10^{-13}$ J, $L = 10^{-15}$ m; $\alpha = 4.4 \times 10^{14}$ m^{-1} and $e^{-2\alpha L} = 0.41$

4)   For $m = 8$ kg, $U - E = 1$ J, $L = 0.02$ m; $\alpha = 3.8 \times 10^{34}$ m^{-1} and $e^{-2\alpha L} = e^{-1.5 \times 10^{33}} \approx 0$

7-7   The continuity requirements from Equation 7.8 are

$$A + B = C + D \qquad \text{[continuity of } \Psi \text{ at } x = 0]$$

$$ikA - ikB = \alpha D - \alpha C \qquad \left[\text{continuity of } \frac{\partial \Psi}{\partial x} \text{ at } x = 0\right]$$

$$Ce^{-\alpha L} + De^{+\alpha L} = Fe^{ikL} \qquad \text{[continuity of } \Psi \text{ at } x = L]$$

$$\alpha De^{+\alpha L} - \alpha Ce^{-\alpha L} = ikFe^{ikL} \qquad \left[\text{continuity of } \frac{\partial \Psi}{\partial x} \text{ at } x = L\right]$$

To isolate the transmission amplitude $\dfrac{F}{A}$, we must eliminate from these relations the unwanted coefficients B, C, and D. Dividing the second line by ik and adding to the first eliminates B, leaving A in terms of C and D. In the same way, dividing the fourth line by α and adding the result to the third line gives D (in terms of F), while subtracting the result from the third line gives C (in terms of F). Combining these results finally yields A:

$$A = \frac{1}{4}Fe^{ikL}\left\{\left[2 - \left(\frac{\alpha}{ik} + \frac{ik}{\alpha}\right)\right]e^{+\alpha L} + \left[2 + \left(\frac{\alpha}{ik} + \frac{ik}{\alpha}\right)\right]e^{-\alpha L}\right\}.$$ The transmission probability is $T = \left|\dfrac{F}{A}\right|^2$.

Making use of the identities $e^{\pm\alpha L} = \cosh\alpha L \pm \sinh\alpha L$ and $\cosh^2\alpha L = 1 + \sinh^2\alpha L$, we obtain

$$\frac{1}{T} = \left|\frac{A}{F}\right|^2 = \frac{1}{4}\left|2\cosh\alpha L + i\left(\frac{\alpha}{k} - \frac{k}{\alpha}\right)\sinh\alpha L\right|^2 = \cosh^2\alpha L + \frac{1}{4}\left(\frac{\alpha}{k} - \frac{k}{\alpha}\right)^2\sinh\alpha L$$

$$= 1 + \frac{1}{4}\left[\frac{U-E}{E} + \frac{E}{U-E} + 2\right]\sinh^2\alpha L = 1 + \frac{1}{4}\left[\frac{U^2}{E(U-E)}\right]\sinh^2\alpha L$$

7-11   (a)   The matter wave reflected from the trailing edge of the well $(x = L)$ must travel the extra distance $2L$ before combining with the wave reflected from the leading edge $(x = 0)$. For $\lambda_2 = 2L$, these two waves interfere destructively since the latter suffers a phase shift of 180° upon reflection, as discussed in Example 7.3.

(b)   The wave functions in all three regions are free particle plane waves. In regions 1 and 3 where $U(x) = U$ we have

$$\Psi(x, t) = Ae^{i(k'x - \omega t)} + Be^{i(-k'x - \omega t)} \qquad x < 0$$
$$\Psi(x, t) = Fe^{i(k'x - \omega t)} + Ge^{i(-k'x - \omega t)} \qquad x < 0$$

with $k' = \dfrac{[2m(E-U)]^{1/2}}{\hbar}$. In this case $G = 0$ since the particle is incident from the left. In region 2 where $U(x) = 0$ we have

$$\Psi(x,\,t) = Ce^{i(-kx-\omega t)} + De^{i(kx-\omega t)} \qquad 0 < x < L$$

with $k = \dfrac{(2mE)^{1/2}}{\hbar} = \dfrac{2\pi}{\lambda_2} = \dfrac{\pi}{L}$ for the case of interest. The wave function and its slope are continuous everywhere, and in particular at the well edges $x = 0$ and $x = L$. Thus, we must require

$A + B = C + D$	[continuity of Ψ at $x = 0$]
$k'A - k'B = kD - kC$	$\left[\text{continuity of } \dfrac{\partial \Psi}{\partial x} \text{ at } x = 0\right]$
$Ce^{-ikL} + De^{ikL} = Fe^{ik'L}$	[continuity of Ψ at $x = L$]
$kDe^{ikL} - kCe^{-ikL} = k'Fe^{ik'L}$	$\left[\text{continuity of } \dfrac{\partial \Psi}{\partial x} \text{ at } x = L\right]$

For $kL = \pi$, $e^{\pm ikL} = -1$ and the last two requirements can be combined to give $kD - kC = k'C + k'D$. Substituting this into the second requirement implies $A - B = C + D$, which is consistent with the first requirement only if $B = 0$, i.e., no reflected wave in region 1.

7-13   As in Problem 7-12, waveform continuity and the slope condition at the site of the delta well demand $A + B = F$ and $ik(A-B) - ikF = -\left(\dfrac{2mS}{\hbar^2}\right)F$. Dividing the second of these equations by ik and subtracting from the first gives $2B + F = F + \dfrac{(2mS/\hbar^2)F}{ik}$, or $B = -i\left(\dfrac{mS}{\hbar^2 k}\right)F = -iF\left(\dfrac{-E_0}{E}\right)^{1/2}$. Thus, the reflection coefficient R is $R(E) = \left|\dfrac{B}{A}\right|^2 = \left|\dfrac{B}{F}\right|^2 \left|\dfrac{F}{A}\right|^2 = \left(\dfrac{-E_0}{E}\right)\left[1 + \left(\dfrac{-E_0}{E}\right)\right]^{-1}$. Then, with $T(E)$ from Problem 7-12, $T(E) = \left[1 + \left(\dfrac{-E_0}{E}\right)\right]^{-1}$, we find $R(E) + T(E) = \left(1 - \dfrac{E_0}{E}\right)\left[1 + \left(\dfrac{-E_0}{E}\right)\right]^{-1} = 1$.

7-15   Divide the barrier region into N subintervals of length $\Delta x = x_{i+1} - x_i$. For the barrier in the ith subinterval, denote by A_i and F_i the incident and transmitted wave amplitudes, respectively. The transmission coefficient for this interval is then $T_i = \left|\dfrac{F_i}{A_i}\right|^2$, and that for the entire barrier is $T(E) = \left|\dfrac{F_N}{A_1}\right|^2$. Now consider the product $\Pi T_i = T_1 T_2 T_3 \ldots T_N = \left(\dfrac{|F_1|^2}{|A_1|^2}\right)\left(\dfrac{|F_2|^2}{|A_2|^2}\right)\left(\dfrac{|F_3|^2}{|A_3|^2}\right)\cdots\left(\dfrac{|F_N|^2}{|A_N|^2}\right)$. Assuming the transmitted wave intensity for one barrier becomes the incident wave intensity for the next, we have $|F_1|^2 = |A_2|^2$, $|F_2|^2 = |A_3|^2$ etc., so that $T(E) = \left|\dfrac{F_N}{A_1}\right|^2 = T_1 T_2 T_3 \ldots T_N$. Next, we assume that Δx is sufficiently small and that $U(x)$ is sensibly constant over each interval (so that the square barrier result can be used for T_i), yet large enough to approximate $\sinh \alpha_i \Delta x$ with $\dfrac{1}{2}e^{\alpha_i \Delta x}$, where α_i, is the value taken by α in the ith subinterval: $\alpha_i = \dfrac{[2m(U_i - E)]^{1/2}}{\hbar}$.

Then, $\dfrac{1}{T_i} = 1 + \left[\dfrac{U_i^2}{4E(U_i - E)}\right]\sinh^2(\alpha_i \Delta x) \approx \left[\dfrac{U_i^2}{16E(U_i - E)}\right]e^{2\alpha_i \Delta x}$ and the transmission coefficient

for the entire barrier becomes $T(E) \approx \Pi\left\{\left[\dfrac{16E(U_i - E)}{U_i^2}\right]e^{-2\alpha_i \Delta x}\right\} \approx \left[\dfrac{\Pi 16E(U_i - E)}{U_i^2}\right]e^{-\Sigma 2\alpha_i \Delta x}$. To

recover Equation 7.10, we approximate the sum in the exponential by an integral, and note

that the product in square brackets is a term of order 1: $T(E) \sim e^{\Sigma 2\alpha_i \Delta x} \approx e^{-\int 2\alpha(x)dx}$ where now

$\alpha(x) = \dfrac{2m[U(x) - E]^{1/2}}{\hbar}$.

7-17   The collision frequency f is the reciprocal of the transit time for the alpha particle crossing the

nucleus, or $f = \dfrac{v}{2R}$, where v is the speed of the alpha. Now v is found from the kinetic energy

which, inside the nucleus, is not the total energy E but the difference $E - U$ between the total

energy and the potential energy representing the bottom of the nuclear well. At the nuclear

radius $R = 9$ fm, the Coulomb energy is

$$\frac{k(Ze)(2e)}{R} = 2Z\left(\frac{ke^2}{a_0}\right)\left(\frac{a_0}{R}\right) = 2(88)(27.2 \text{ eV})\left(\frac{5.29 \times 10^4 \text{ fm}}{9 \text{ fm}}\right) = 28.14 \text{ MeV}.$$

From this we conclude that $U = -1.86$ MeV to give a nuclear barrier of 30 MeV overall. Thus

an alpha with $E = 4.05$ MeV has kinetic energy $4.05 + 1.86 = 5.91$ MeV inside the nucleus. Since

the alpha particle has the combined mass of 2 protons and 2 neutrons, or about

$3\,755.8 \text{ MeV}/c^2$ this kinetic energy represents a speed

$$v = \left(\frac{2E_k}{m}\right)^{1/2} = \left[\frac{2(5.91)}{3\,755.8 \text{ MeV}/c^2}\right]^{1/2} = 0.056c.$$

Thus, we find for the collision frequency $f = \dfrac{v}{2R} = \dfrac{0.056c}{2(9 \text{ fm})} = 9.35 \times 10^{20}$ Hz.

8

Quantum Mechanics in Three Dimensions

8-1 $E = \dfrac{\hbar^2 \pi^2}{2m} \left[\left(\dfrac{n_1}{L_x} \right)^2 + \left(\dfrac{n_2}{L_y} \right)^2 + \left(\dfrac{n_3}{L_z} \right)^2 \right]$

$L_x = L$, $L_y = L_z = 2L$. Let $\dfrac{\hbar^2 \pi^2}{8mL^2} = E_0$. Then $E = E_0 \left(4n_1^2 + n_2^2 + n_3^2 \right)$. Choose the quantum numbers as follows:

n_1	n_2	n_3	$\dfrac{E}{E_0}$		
1	1	1	6		ground state
1	2	1	9	*	first two excited states
1	1	2	9	*	
2	1	1	18		
1	2	2	12	*	next excited state
2	1	2	21		
2	2	1	21		
2	2	2	24		
1	1	3	14	*	next two excited states
1	3	1	14	*	

Therefore the first 6 states are ψ_{111}, ψ_{121}, ψ_{112}, ψ_{122}, ψ_{113}, and ψ_{131} with relative energies $\dfrac{E}{E_0} = 6, 9, 9, 12, 14, 14$. First and third excited states are doubly degenerate.

8-3 $n^2 = 11$

(a) $E = \left(\dfrac{\hbar^2 \pi^2}{2mL^2} \right) n^2 = \dfrac{11}{2} \left(\dfrac{\hbar^2 \pi^2}{mL^2} \right)$

(b)

n_1	n_2	n_3	
1	1	3	
1	3	1	3-fold degenerate
3	1	1	

53

(c) $\psi_{113} = A\sin\left(\dfrac{\pi x}{L}\right)\sin\left(\dfrac{\pi y}{L}\right)\sin\left(\dfrac{3\pi z}{L}\right)$

$\psi_{131} = A\sin\left(\dfrac{\pi x}{L}\right)\sin\left(\dfrac{3\pi y}{L}\right)\sin\left(\dfrac{\pi z}{L}\right)$

$\psi_{311} = A\sin\left(\dfrac{3\pi x}{L}\right)\sin\left(\dfrac{\pi y}{L}\right)\sin\left(\dfrac{\pi z}{L}\right)$

8-5   (a) $n_1 = n_2 = n_3 = 1$ and $E_{111} = \dfrac{3h^2}{8mL^2} = \dfrac{3(6.63\times10^{-34})^2}{8(1.67\times10^{-27})(4\times10^{-28})} = 2.47\times10^{-13}$ J ≈ 1.54 MeV

(b)   States 211, 121, 112 have the same energy and $E = \dfrac{(2^2 + 1^2 + 1^2)h^2}{8mL^2} = 2E_{111} \approx 3.08$ MeV

and states 221, 122, 212 have the energy $E = \dfrac{(2^2 + 2^2 + 1^2)h^2}{8mL^2} = 3E_{111} \approx 4.63$ MeV.

(c)   Both states are threefold degenerate.

8-7   The stationary states for a particle in a cubic box are, from Equation 8.10

$$\Psi(x,\, y,\, z,\, t) = A\sin(k_1 x)\sin(k_2 y)\sin(k_3 z)e^{-iEt/\hbar} \quad 0 \le x,\, y,\, x \le L$$
$$= 0 \text{ elsewhere}$$

where $k_1 = \dfrac{n_1 \pi}{L}$, etc. Since Ψ is nonzero only for $0 < x < L$, and so on, the normalization condition reduces to an integral over the volume of a cube with one corner at the origin:

$$1 = \int dx \int dy \int dz |\Psi(\mathbf{r},\, t)|^2 = A^2 \left\{ \int_0^L \sin^2(k_1 x)dx \int_0^L \sin^2(k_2 y)dy \int_0^L \sin^2(k_3 z)dz \right\}$$

Using $2\sin^2\theta = 1 - \cos 2\theta$ gives $\int_0^L \sin^2(k_1 x)dx = \dfrac{L}{2} - \dfrac{1}{4k_1}\sin(2k_1 x)\Big|_0^L$. But $k_1 L = n_1\pi$, so the last term on the right is zero. The same result is obtained for the integrations over y and z. Thus, normalization requires $1 = A^2\left(\dfrac{L}{2}\right)^3$ or $A = \left(\dfrac{2}{L}\right)^{3/2}$ for any of the stationary states. Allowing the edge lengths to be different at L_1, L_2, and L_3 requires only that L^3 be replaced by the box volume $L_1 L_2 L_3$ in the final result: $A = \left\{\left(\dfrac{2}{L_1}\right)\left(\dfrac{2}{L_2}\right)\left(\dfrac{2}{L_3}\right)\right\}^{1/2} = \left(\dfrac{8}{L_1 L_2 L_3}\right)^{1/2} = \left(\dfrac{8}{V}\right)^{1/2}$ where $V = L_1 L_2 L_3$ is the volume of the box. This follows because it is still true that the wave must vanish at the walls of the box, so that $k_1 L_1 = n_1\pi$, and so on.

8-9 $L = [l(l+1)]^{1/2}\hbar$

4.714×10^{-34} Js $= [l(l+1)]^{1/2}\left(\dfrac{6.63\times10^{-34} \text{ Js}}{2\pi}\right)$

$l(l+1) = \dfrac{(4.714\times10^{-34})^2(2\pi)^2}{(6.63\times10^{-34})^2} = 1.996\times10^1 \approx 20 = 4(4+1)$

so $l = 4$.

8-11   (a) $L = [l(l+1)]^{1/2}\hbar$; $\ 4.83 \times 10^{31}$ Js $= [l(l+1)]^{1/2}\hbar$, so

$$l^2 + l = \frac{\left(4.83 \times 10^{31} \text{ Js}\right)^2}{\left(1.055 \times 10^{-34} \text{ Js}\right)^2} \approx \left(4.58 \times 10^{65}\right)^2 \approx l^2$$

$$l \approx 4.58 \times 10^{65}$$

(b)   With $L \approx l\hbar$ we get $\Delta L \approx \hbar$ and $\dfrac{\Delta L}{L} \approx \dfrac{1}{l} = 2.18 \times 10^{-66}$

8-13 $Z = 2$ for He$^+$

(a)   For $n = 3$, l can have the values of 0, 1, 2

$$\begin{aligned} l &= 0 &\rightarrow&\quad m_l = 0 \\ l &= 1 &\rightarrow&\quad m_l = -1,\, 0,\, +1 \\ l &= 2 &\rightarrow&\quad m_l = -2,\, -1,\, 0,\, +1,\, +2 \end{aligned}$$

(b)   All states have energy $E_3 = \dfrac{-Z^2}{3^2}(13.6 \text{ eV})$

$$E_3 = -6.04 \text{ eV}.$$

8-15   (a) $E_n = -\left(\dfrac{ke^2}{2a_0}\right)\left(\dfrac{Z^2}{n^2}\right)$ from Equation 8.38. But $a_0 = \dfrac{\hbar^2}{m_e ke^2}$ so with $m_e \rightarrow \mu$ we get

$E_n = -\left(\dfrac{\mu k^2 e^4}{2\hbar^2}\right)\left(\dfrac{Z^2}{n^2}\right)$.

(b)   For $n = 3 \rightarrow 2$, $E_3 - E_2 = \dfrac{hc}{\lambda} = \dfrac{\mu k^2 e^4 Z^2}{2\hbar^2}\left(\dfrac{1}{2^2} - \dfrac{1}{3^2}\right)$ with $\lambda = 656.3$ nm for H ($Z = 1$,

$\mu \approx m_e$). For He$^+$, $Z = 2$, and $\mu \approx m_e$, so, $\lambda = \dfrac{656.3}{2^2} = 164.1$ nm (ultraviolet).

(c)   For positronium, $Z = 1$ and $\mu = \dfrac{m_e}{2}$, so, $\lambda = (656.3)(2) = 1\,312.6$ nm (infrared).

8-17   (a)   For a d state, $l = 2$

$$L = [l(l+1)]^{1/2}\hbar = (6)^{1/2}\left(1.055 \times 10^{-34} \text{ Js}\right) = 2.58 \times 10^{-34} \text{ Js}$$

(b)   For an f state, $l = 3$

$$L = [l(l+1)]^{1/2}\hbar = (12)^{1/2}\left(1.055 \times 10^{-34} \text{ Js}\right) = 3.65 \times 10^{-34} \text{ Js}$$

8-19   When the principal quantum number is n, the following values of l are possible: $l = 0, 1, 2, \ldots, n-2, n-1$. For a given value of l, there are $2l+1$ possible values of m_l. The maximum number of electrons that can be accommodated in the nth level is therefore:

$$(2(0)+1)+(2(1)+1)+\ldots+(2l+1)+\ldots+(2(n-1)+1) = 2\sum_{l=0}^{n-1} l + \sum_{l=0}^{n-1} l = 2\sum_{l=0}^{n-1} l + n.$$

But $\sum_{l=0}^{k} l = \dfrac{k(k+1)}{2}$ so the maximum number of electrons to be accommodated is

$$\frac{2(n-1)n}{2} + n = n^2.$$

8-21   (a) $\psi_{2s}(r) = \dfrac{1}{4(2\pi)^{1/2}} \left(\dfrac{1}{a_0}\right)^{3/2} \left(2 - \dfrac{r}{a_0}\right) e^{-r/2a_0}$. At $r = a_0 = 0.529 \times 10^{-10}$ m we find

$$\psi_{2s}(a_0) = \frac{1}{4(2\pi)^{1/2}} \left(\frac{1}{a_0}\right)^{3/2} (2-1)e^{-1/2} = (0.380)\left(\frac{1}{a_0}\right)^{3/2}$$

$$= (0.380)\left[\frac{1}{0.529 \times 10^{-10} \text{ m}}\right]^{3/2} = 9.88 \times 10^{14} \text{ m}^{-3/2}$$

(b) $|\psi_{2s}(a_0)|^2 = \left(9.88 \times 10^{14} \text{ m}^{-3/2}\right)^2 = 9.75 \times 10^{29} \text{ m}^{-3}$

(c)   Using the result to part (b), we get $P_{2s}(a_0) = 4\pi a_0^2 |\psi_{2s}(a_0)|^2 = 3.43 \times 10^{10} \text{ m}^{-1}$.

8-23   (a) $\dfrac{1}{\alpha} = \dfrac{\hbar c}{ke^2} = \dfrac{\left(6.63 \times 10^{-34} \text{ Js}\right)\left(3 \times 10^8 \text{ m/s}\right)}{2\pi \left(9 \times 10^9 \text{ N m}^2/\text{C}^2\right)\left(1.6 \times 10^{-19} \text{ C}\right)^2} = 137.036$

(b) $\dfrac{\lambda_c}{r_0} = \dfrac{h/m_e c}{ke^2/m_e c^2} = \dfrac{hc}{ke^2} = \dfrac{2\pi}{\alpha} = 2\pi \times 137$

(c) $\dfrac{a_0}{\lambda_c} = \dfrac{\hbar^2/m_e ke^2}{h/m_e c} = \dfrac{1}{2\pi}\dfrac{\hbar c}{ke^2} = \dfrac{1}{2\pi\alpha} = \dfrac{137}{2\pi}$

(d) $\dfrac{1}{Ra_0} = \left(\dfrac{m_e ke^2}{\hbar^2}\right)\left(\dfrac{4\pi c\hbar^3}{m_e k^2 e^4}\right) = \dfrac{4\pi\hbar c}{ke^2} = \dfrac{4\pi}{\alpha} = 4\pi(137)$

8-25   The most probable distance is the value of r which maximizes the radial probability density $P(r) = |rR(r)|^2$. Since $P(r)$ is largest where $rR(r)$ reaches its maximum, we look for the most probable distance by setting $\dfrac{d\{rR(r)\}}{dr}$ equal to zero, using the functions $R(r)$ from Table 8.4. For clarity, we measure distances in bohrs, so that $\dfrac{r}{a_0}$ becomes simply r, etc. Then for the $2s$ state of hydrogen, the condition for a maximum is

$$0 = \frac{d}{dr}\left\{(2r - r^2)e^{-r/2}\right\} = \left\{2 - 2r - \frac{1}{2}(2r - r^2)\right\}e^{-r/2}$$

MODERN PHYSICS   57

or $0 = 4 - 6r + r^2$. There are two solutions, which may be found by completing the square to get $0 = (r-3)^2 - 5$ or $r = 3 \pm \sqrt{5}$ bohrs. Of these $r = 3 + \sqrt{5} = 5.236a_0$ gives the largest value of $P(r)$, and so is the most probable distance. For the $2p$ state of hydrogen, a similar analysis gives $0 = \frac{d}{dr}\{r^2 e^{-r/2}\} = \{2r - \frac{1}{2}r^2\}e^{-r/2}$ with the obvious roots $r = 0$ (a minimum) and $r = 4$ (a maximum). Thus, the most probable distance for the $2p$ state is $r = 4a_0$, in agreement with the simple Bohr model.

8-29   To find Δr we first compute $\langle r^2 \rangle$ using the radial probability density for the $1s$ state of hydrogen: $P_{1s}(r) = \frac{4}{a_0^3}r^2 e^{-2r/a_0}$. Then $\langle r^2 \rangle = \int_0^\infty r^2 P_{1s}(r)dr = \frac{4}{a_0^3}\int_0^\infty r^4 e^{-2r/a_0}dr$. With $z = \frac{2r}{a_0}$, this is

$\langle r^2 \rangle = \frac{4}{a_0^3}\left(\frac{a_0}{2}\right)^5 \int_0^\infty z^4 e^{-z}dz$. The integral on the right is (see Example 8.9) $\int_0^\infty z^4 e^{-z}dz = 4!$ so that

$\langle r^2 \rangle = \frac{4}{a_0^3}\left(\frac{a_0}{2}\right)^5 (4!) = 3a_0^2$ and $\Delta r = (\langle r^2 \rangle - \langle r \rangle^2)^{1/2} = [3a_0^2 - (1.5a_0)^2]^{1/2} = 0.866a_0$. Since Δr is an appreciable fraction of the average distance, the whereabouts of the electron are largely unknown in this case.

8-31   Outside the surface, $U(x) = -\frac{A}{x}$ (to give $F = -\frac{dU}{dx} = -\frac{A}{x^2}$), and Schrödinger's equation is

$-\left(\frac{\hbar^2}{2m_e}\right)\frac{d^2\psi}{dx^2} + \left(-\frac{A}{x}\right)\psi(x) = E\psi(x)$. From Equation 8.36 $g(r) = rR(r)$ satisfies a one-dimensional

Schrödinger equation with effective potential $U_{eff}(r) = U(r) + \frac{l(l+1)\hbar^2}{2m_e r^2}$. With $l = 0$ (s states)

and $U(r) = -\frac{kZe^2}{r}$ the equation for $g(r)$ has the same form as that for $\psi(x)$. Furthermore, $\psi(0) = 0$ if no electrons can cross the surface, while $g(0) = 0$ since $R(0)$ must be finite. It follows that the functions $g(r)$ and $\psi(x)$ are the same, and that the energies in the present case are the hydrogenic levels $E_n = -\left(\frac{Z^2 ke^2}{2a_0}\right)\left(\frac{1}{n^2}\right)$ with the replacement $kZe^2 \rightarrow A$.

Remembering that $a_0 = \frac{\hbar^2}{m_e ke^2}$, we get $E_n = -\left(\frac{mA^2}{2\hbar^2}\right)\left(\frac{1}{n^2}\right)$ $n = 1, 2, \ldots$

8-34   (a)   Probability $= \int_r^\infty P_{1s}(r')dr' = \frac{4}{a_0^3}\int_r^\infty r'^2 e^{-2r'/a_0}dr' = \left[-\left(\frac{2r'^2}{a_0^2} + \frac{2r'}{a_0} + 1\right)e^{-2r'/a_0}\right]_r^\infty$,

using integration by parts

$$= \boxed{\left(\frac{2r^2}{a_0^2} + \frac{2r}{a_0} + 1\right)e^{-2r/a_0}}$$

(b)   The probability of finding the electron inside or outside the sphere of radius r is $\frac{1}{2}$.

$\therefore \left(\frac{2r^2}{a_0^2} + \frac{2r}{a_0} + 1\right)e^{-2r/a_0} = \frac{1}{2}$ or $z^2 + 2z + 2 = e^z$ where $z = \frac{2r}{a_0}$

One can home in on a solution to this transcendental equation for r on a calculator, the result being $r = \boxed{1.34a_0}$ to three digits.

9

Atomic Structure

9-1 $\Delta E = 2\mu_B B = hf$

$2(9.27 \times 10^{-24} \text{ J/T})(0.35 \text{ T}) = (6.63 \times 10^{-34} \text{ Js})f$ so $f = 9.79 \times 10^9$ Hz

9-3   (a) $n = 1$; for $n = 1$, $l = 0$, $m_l = 0$, $m_s = \pm\dfrac{1}{2}$ $\rightarrow$ 2 sets

n	l	m_l	m_s
1	0	0	−1/2
1	0	0	+1/2

$2n^2 = 2(1)^2 = 2$

(b)   For $n = 2$ we have

n	l	m_l	m_s
2	0	0	±1/2
2	1	−1	±1/2
2	1	0	±1/2
2	1	1	±1/2

Yields 8 sets; $2n^2 = 2(2)^2 = 8$. Note that the number is twice the number of m_l values. Also that for each l there are $2l + 1$ m_l values. Finally, l can take on values ranging from 0 to $n - 1$, so the general expression is $s = \sum_{0}^{n-1} 2(2l + 1)$. The series is an arithmetic progression: $2 + 6 + 10 + 14\ldots$, the sum of which is

$$s = \frac{n}{2}[2a + (n - 1)d] \qquad \text{where } a = 2,\ d = 4$$

$$s = \frac{n}{2}[4 + (n - 1)4] = 2n^2$$

(c) $n = 3$: $2(1) + 2(3) + 2(5) = 2 + 6 + 10 = 18 = 2n^2 = 2(3)^2 = 18$

(d) $n = 4$: $2(1) + 2(3) + 2(5) + 2(7) = 32 = 2n^2 = 2(4)^2 = 32$

(e) $n = 5$: $32 + 2(9) = 32 + 18 = 50 = 2n^2 = 2(5)^2 = 50$

9-5   The time of passage is $t = \dfrac{1 \text{ m}}{100 \text{ m/s}} = 0.01$ s. Since the field gradient is assumed uniform, so is

the force, and hence the acceleration. Thus the deflection is $d = \dfrac{1}{2}at^2$, or $a = \dfrac{2d}{t^2}$ for the

acceleration. The required force is then

$$F_z = \frac{M2d}{t^2} = \frac{2(108 \text{ u})(1.66 \times 10^{-27} \text{ kg/u})(10^{-3} \text{ m})}{(10^{-2} \text{ s})^2} = 3.59 \times 10^{-24} \text{ N}.$$

The magnetic moment of the silver atom is due to a single unpaired electron spin, so

$$\mu_z = 2\left(\frac{e}{2m_e}\right)S_z = 2\left(\frac{e}{2m_e}\right)\left(\frac{\hbar}{2}\right) = \mu_B = 9.27 \times 10^{-24} \text{ J/T}.$$

Thus,

$$\frac{dB_z}{dz} = \frac{F_z}{\mu_z} = \frac{3.59 \times 10^{-24} \text{ N}}{9.27 \times 10^{-24} \text{ N}} = 0.387 \text{ T/m}.$$

9-7   The angular momentum **L** of a spinning ball is related to the angular velocity of rotation **ω** as
L = I**ω**. I, the moment of inertia, is given in terms of the mass m and radius R of the ball as
$I = \dfrac{2}{5}mR^2$. For the electron this gives

$$I = \frac{2}{5}(511 \times 10^3 \text{ eV}/c^2)(3 \times 10^{-6} \text{ nm})^2 = 1.840 \times 10^{-6} \text{ eV nm}^2/c^2.$$

Then, using $L = \dfrac{\sqrt{3}}{2}\hbar$, we find $\omega = \dfrac{L}{I} = \dfrac{\sqrt{3}}{2}\dfrac{(197.3 \text{ eV nm}/c)}{1.840 \times 10^{-6} \text{ eV nm}^2/c^2} = 9.286 \times 10^7$ c/nm. The

equatorial speed is

$$v = R\omega = (3 \times 10^{-6} \text{ nm})(9.286 \times 10^7 \text{ } c/\text{nm}) = 278.6c$$

$$\frac{v}{c} = 278.6$$

9-9   With $s = \dfrac{3}{2}$, the spin magnitude is $|\mathbf{S}| = [s(s+1)]^{1/2}\hbar = \left(\dfrac{[15]^{1/2}}{2}\right)\hbar$. The z-component of spin is

$S_z = m_s\hbar$ where m_s ranges from $-s$ to s in integer steps or, in this case,
$m_s = -\dfrac{3}{2}, -\dfrac{1}{2}, +\dfrac{1}{2}, +\dfrac{3}{2}$. The spin vector S is inclined to the z-axis by an angle θ such that

$$\cos(\theta) = \frac{S_z}{|\mathbf{S}|} = \frac{m_s\hbar}{([15]^{1/2}/2)\hbar} = \frac{m_s}{[15]^{1/2}/2} = -\frac{3}{(15)^{1/2}}, -\frac{1}{(15)^{1/2}}, +\frac{1}{(15)^{1/2}}, +\frac{3}{(15)^{1/2}}$$

or $\theta = 140.8°$, $105.0°$, $75.0°$, $39.2°$. The Ω^- *does* obey the Pauli Exclusion Principle, since the
spin s of this particle is half-integral, as it is for all fermions.

9-11   For a d electron, $l = 2$; $s = \frac{1}{2}$; $j = 2 + \frac{1}{2}, 2 - \frac{1}{2}$

For $j = \frac{5}{2}$; $m_j = -\frac{5}{2}, -\frac{3}{2}, -\frac{1}{2}, \frac{1}{2}, \frac{3}{2}, \frac{5}{2}$

For $j = \frac{3}{2}$; $m_j = -\frac{3}{2}, -\frac{1}{2}, \frac{1}{2}, \frac{3}{2}$

9-13   (a) $4F_{5/2} \rightarrow n = 4, l = 3, j = \frac{5}{2}$

(b) $|\mathbf{J}| = [j(j+1)]^{1/2} \hbar = \left[\left(\frac{5}{2} \right) \left(\frac{7}{2} \right) \right]^{1/2} \hbar = \left[\frac{35}{4} \right]^{1/2} \hbar = \left[\frac{(35)^{1/2}}{2} \right] \hbar$

(c) $J_z = m_j \hbar$ where m_j can be $-j, -j+1, \ldots, j-1, j$ so here m_j can be

$-\frac{5}{2}, -\frac{3}{2}, -\frac{1}{2}, \frac{1}{2}, \frac{3}{2}, \frac{5}{2}$. J_z can be $-\frac{5}{2}\hbar, -\frac{3}{2}\hbar, -\frac{1}{2}\hbar, \frac{1}{2}\hbar, \frac{3}{2}\hbar$, or $\frac{5}{2}\hbar$.

9-15   The spin of the atomic electron has a magnetic energy in the field of the orbital moment given

by Equations 9.6 and 9.12 with a g-factor of 2, or $U = -\boldsymbol{\mu}_s \cdot \mathbf{B} = 2 \left(\frac{e}{2m_e} \right) S_z B = 2\mu_B m_s B$. The

magnetic field $\mathbf{B}$ originates with the orbiting electron. To estimate $\mathbf{B}$, we adopt the equivalent
viewpoint of the atomic nucleus (proton) circling the electron, and borrow a result from
classical electromagnetism for the $\mathbf{B}$ field at the center of a circular current loop with radius r:

$B = \frac{2k_m \mu}{r^3}$. Here k_m is the magnetic constant and $\mu = i\pi r^2$ is the magnetic moment of the

loop, assuming it carries a current i. In the atomic case, we identify r with the orbit radius and

the current i with the proton charge $+e$ divided by the orbital period $T = \frac{2\pi r}{v}$. Then

$\mu = \frac{evr}{2} = \left(\frac{e}{2m_e} \right) L$ where $L = m_e vr$ is the orbital angular momentum of the *electron*. For a p

electron $l = 1$ and $L = [l(l+1)]^{1/2} \hbar = \sqrt{2}\hbar$, so $\mu = \left(\frac{e\hbar}{2m_e} \right) \sqrt{2} = \mu_B \sqrt{2} = 1.31 \times 10^{-23}$ J/T. For r we

take a typical atomic dimension, say $4a_0 \left(= 2.12 \times 10^{-10} \text{ m} \right)$ for a $2p$ electron, and find

$$B = \frac{2 \left(10^{-7} \text{ N/A}^2 \right) \left(1.31 \times 10^{-23} \text{ J/T} \right)}{\left(2.12 \times 10^{-10} \text{ m} \right)^3} = 0.276 \text{ T}.$$

Since m_s is $\pm \frac{1}{2}$ the magnetic energy of the electron spin in this field is

$$U = \pm \mu_B B = \pm \left(9.27 \times 10^{-24} \text{ J/T} \right) (0.276 \text{ T}) = \pm 2.56 \times 10^{-24} \text{ J} = \pm 1.59 \times 10^{-5} \text{ eV}.$$

The up spin orientation (+) has the higher energy; the predicted energy difference between
the up (+) and down (−) spin orientations is twice this figure, or about 3.18×10^{-5} eV —a
result which compares favorably with the measured value, 5×10^{-5} eV.

9-17   From Equation 8.9 we have $E = \left(\dfrac{\hbar^2 \pi^2}{2mL^2}\right)\left(n_1^2 + n_2^2 + n_3^2\right)$

$$E = \frac{\left(1.054 \times 10^{-34}\right)^2 \left(\pi^2\right)\left(n_1^2 + n_2^2 + n_3^2\right)}{2\left(9.11 \times 10^{-31}\right)\left(2 \times 10^{-10}\right)^2} = \left(1.5 \times 10^{-18} \text{ J}\right)\left(n_1^2 + n_2^2 + n_3^2\right) = \left(9.4 \text{ eV}\right)\left(n_1^2 + n_2^2 + n_3^2\right)$$

(a)   2 electrons per state. The lowest states have

$$\left(n_1^2 + n_2^2 + n_3^2\right) = (1,\,1,\,1) \Rightarrow E_{111} = (9.4 \text{ eV})\left(1^2 + 1^2 + 1^2\right) \text{ eV} = 28.2 \text{ eV}.$$

For $\left(n_1^2 + n_2^2 + n_3^2\right) = (1,\,1,\,2)$ or $(1,\,2,\,1)$ or $(2,1,1)$,

$$E_{112} = E_{121} = E_{211} = (9.4 \text{ eV})\left(1^2 + 1^2 + 2^2\right) = 56.4 \text{ eV}$$
$$E_{\min} = 2 \times \left(E_{111} + E_{112} + E_{121} + E_{211}\right) = 2(28.2 + 3 \times 56.4) = 398.4 \text{ eV}$$

(b)   All 8 particles go into the $\left(n_1^2 + n_2^2 + n_3^2\right) = (1,\,1,\,1)$ state, so

$$E_{\min} = 8 \times E_{111} = 225.6 \text{ eV}.$$

9-21   (a) $1s^2 2s^2 2p^4$

(b)   For the two 1s electrons, $n = 1, \quad l = 0, \ m_l = 0, \ m_s = \pm\dfrac{1}{2}$.

For the two 2s electrons, $n = 2, \ l = 0, \ m_l = 0, \ m_s = \pm\dfrac{1}{2}$.

For the four 2p electrons, $n = 2, \ l = 1, \ m_l = 1,\ 0,\ -1, \ m_s = \pm\dfrac{1}{2}$.

9-23   All spins are paired for $[\text{Kr}]4d^{10}$ and two are unpaired for $[\text{Kr}]4d^9 5s^1$. Thus Hund's rule would favor the latter, but for the fact that completely filled subshells are especially stable. Thus $[\text{Kr}]4d^{10}$ with its completely filled 4d subshell has the lesser energy. The element is palladium (Pd).

9-25   A typical ionization energy is 8 eV. For internal energy to ionize most of the atoms would require $\dfrac{3}{2}k_B T = 8 \text{ eV} : T = \dfrac{2 \times 8\left(1.60 \times 10^{-19} \text{ J}\right)}{3\left(1.38 \times 10^{-23} \text{ J/K}\right)} \sim$ between 10^4 K and 10^5 K.

9-27   (a)   The L_α photon can be thought of as arising from the $n = 3$ to $n = 2$ transition in a one-electron atom with an effective nuclear charge. The M electron making the transition is shielded by the remaining L shell electrons (5) and the innermost K shell electrons (2), leaving an effective nuclear charge of $Z - 7$. Thus, the energy of the L_α photon should be $E[L_\alpha] = \dfrac{ke^2}{2a_0}\dfrac{(Z-7)^2}{3^2} + \dfrac{ke^2}{2a_0}\dfrac{(Z-7)^2}{2^2} + \dfrac{ke^2}{2a_0}\dfrac{5(Z-7)^2}{36}$. Writing $E = hf$ and

noting that $\dfrac{ke^2}{2a_0} = 13.6$ eV this relation may be solved for the photon frequency f.

Taking the square root of the resulting equation gives $\sqrt{f} = \sqrt{\dfrac{5}{36}\left(\dfrac{13.6 \text{ eV}}{h}\right)}(Z - 7)$.

(b)   According to part (a), the plot of $\sqrt{f}$ against Z should have intercept $= 7$ and slope

$$\sqrt{\frac{5}{36}\left(\frac{13.6\ \text{eV}}{h}\right)} = \sqrt{\frac{5(13.6\ \text{eV})}{36(4.14\times 10^{-15}\ \text{eV s})}} = 0.214\times 10^{8}\ \text{Hz}^{1/2}.$$ From Figure 9.18 we find

data points on the L_{α} line [in the form $(\sqrt{f}, Z)$] at $(14, 74)$ and $(8, 45)$. From this we

obtain the slope $\dfrac{14-8}{74-45} = 0.21\times 10^{8}\ \text{Hz}^{1/2}$. Thus, the empirical line fitting the L_{α} data

is $\sqrt{f} = 0.21(Z - I)$ where I is the intercept. Using $(14, 74)$ for $(\sqrt{f}, Z)$ in this equation

gives the intercept $I = 7.3$, but with $(8, 45)$ for $(\sqrt{f}, Z)$ we get $I = 6.9$. Alternatively,

using *both* data pairs and dividing, we eliminate the calculated value of the slope to

get $\dfrac{14}{8} = \dfrac{74 - I}{45 - I}$. This last approach affords the best experimental value for I based on

the available data and gives $I = \dfrac{(14)(45) - (8)(74)}{14 - 8} = 6.3$.

(c)   The average screened nuclear charge seen by the M shell electron is just
$Z - I = Z - 6.3$, indicating that shielding by the inner shell electrons is not quite as
effective as our naïve screening arguments would suggest.

10

Statistical Physics

10-1 Using $\bar{n}_j = n_{j1}p_1 + n_{j2}p_2 + \dots$ we obtain:

$$\bar{n}_1 = n_{11}p_1 + n_{12}p_2 + n_{120}p_{20} = (0)\left(\frac{6}{1\,287}\right) + (1)\left(\frac{30}{1\,287}\right) + (0)\left(\frac{30}{1\,287}\right) + (2)\left(\frac{60}{1\,287}\right) + (0)\left(\frac{30}{1\,287}\right)$$

$$+ (1)\left(\frac{120}{1\,287}\right) + (3)\left(\frac{60}{1\,287}\right) + (0)\left(\frac{15}{1\,287}\right) + (1)\left(\frac{120}{1\,287}\right) + (0)\left(\frac{60}{1\,287}\right) + (2)\left(\frac{180}{1\,287}\right) + (4)\left(\frac{30}{1\,287}\right)$$

$$+ (0)\left(\frac{60}{1\,287}\right) + (2)\left(\frac{90}{1\,287}\right) + (1)\left(\frac{180}{1\,287}\right) + (3)\left(\frac{120}{1\,287}\right) + (5)\left(\frac{6}{1\,287}\right) + (0)\left(\frac{15}{1\,287}\right) + (2)\left(\frac{60}{1\,287}\right)$$

$$+ (4)\left(\frac{15}{1\,287}\right)$$

$$= \frac{30 + 120 + 120 + 180 + 120 + 360 + 120 + 180 + 180 + 360 + 30 + 120 + 60}{1\,287}$$

$$= 1.538\,46$$

$$p(1E) = \frac{\bar{n}_1}{6} = \frac{1.538}{6} = 0.256.$$

One can find $p(2E)$ through $p(8E)$ in similar fashion.

10-3 A molecule moving with speed v takes $\dfrac{d}{v}$ seconds to cross the cylinder, where d is the cylinder's diameter. In this time the detector rotates θ radians where $\theta = \omega t = \dfrac{\omega d}{v}$. This means the molecule strikes the curved glass plate at a distance from A of $s = \dfrac{d}{2}\theta = \dfrac{\omega d^2}{2v}$ as $m_{\mathrm{Bi}_2} = 6.94 \times 10^{-22}$ g and

$$\langle v \rangle = \left[\frac{8k_{\mathrm{B}}T}{\pi m}\right]^{1/2} = \left[\frac{(8)(1.38 \times 10^{-23}\ \mathrm{J/K})(850\ \mathrm{K})}{(\pi)(6.94 \times 10^{-25}\ \mathrm{kg})}\right] = 207\ \mathrm{m/s}$$

$$v_{\mathrm{rms}} = \left(\frac{3k_{\mathrm{B}}T}{m}\right)^{1/2} = 225\ \mathrm{m/s} \qquad v_{\mathrm{mp}} = \left(\frac{2k_{\mathrm{B}}T}{m}\right)^{1/2} = 184\ \mathrm{m/s}$$

$$s_{\mathrm{rms}} = \left(\frac{6\,250 \times 2\pi}{60\ \mathrm{s}}\right)\frac{(0.10\ \mathrm{m})^2}{(2)(225)\ \mathrm{m/s}} = 1.45\ \mathrm{cm}$$

$$s_{\langle v \rangle} = 1.58\ \mathrm{cm} \qquad\qquad s_{\mathrm{mp}} = 1.78\ \mathrm{cm}$$

10-5   Fit a curve Ae^{-BE} to Figure 10.2. An ambitious solution would use a least squares fit to determine A and B. The quick fit suggested below uses a match only at 0 and 1E. $P(E) = Ae^{-BE}$ thus $P(0) = A$ and $P(E_1) = Ae^{-BE_1}$. From Figure 10.2 one finds $P(0) = 0.385$, and this gives $A = 0.385$. To determine B use the value $P(1E) = 0.256 = Ae^{-BE_1} = 0.385e^{-BE_1}$ thus $e^{-BE_1} = 0.665$ and $B = -\dfrac{\ln(0.665)}{E_1} = \dfrac{0.408}{E_1}$ and so $P(E) = (0.385)e^{-(0.408E/E_1)}$. This equation was used to determine the probability as follows $P(0) = 0.385$, $P(1E_1) = 0.256$, $P(2E_1) = 0.170$, $P(3E_1) = 0.113$, $P(4E_1) = 0.075$, $P(5E_1) = 0.050$, $P(6E_1) = 0.033$, $P(7E_1) = 0.022$, $P(8E_1) = 0.015$.

The exact values are $P(0) = 0.385$, $P(1E_1) = 0.256$, $P(2E_1) = 0.167$, $P(3E_1) = 0.078$, $P(4E_1) = 0.054$, $P(5E_1) = 0.027$, $P(6E_1) = 0.012$, $P(7E_1) = 0.003\,9$, $P(8E_1) = 0.000\,717$. These values are plotted below. One sees that this approximation is good for low energy. There is exact agreement for $P(0)$ and $P(1E)$ and small deviations for the next two values with percent deviations for the higher energy values.

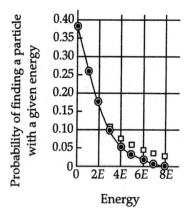

10-7   (a)

$$E_U = -\mathbf{p} \cdot \boldsymbol{\varepsilon} = -\varepsilon\cos 0° = -p\varepsilon$$
$$E_D = -\mathbf{p} \cdot \boldsymbol{\varepsilon} = -\varepsilon\cos 180° = +p\varepsilon$$

so $\Delta E = E_D - E_U = 2p\varepsilon$.

(b)   Let $n(2p\varepsilon)$ be the number of molecules in the excited state.

$$\frac{n(2p\varepsilon)}{n(0)} = \frac{g(2p\varepsilon)Ae^{-2p\varepsilon/k_B T}}{g(0)Ae^0} = 2e^{-2p\varepsilon/k_B T}$$

(c) $\dfrac{1.90}{1} = \dfrac{n(2p\varepsilon)}{n(0)} = 2e^{-2p\varepsilon/k_B T}$. For $p = 1.0 \times 10^{-30}$ Cm and $\varepsilon = \left(1.0 \times 10^6 \text{ V/m}\right)$,

$$\frac{2p\varepsilon}{k_B T} = \frac{(2)\left(1.0 \times 10^{-30} \text{ Cm}\right)\left(1.0 \times 10^6 \text{ V/m}\right)}{\left(1.38 \times 10^{-23} \text{ J/K}\right)T} = \frac{0.1449}{T}$$

so $1.90 = 2e^{-0.1449/T}$ or $0.95 = e^{-0.1449/T}$. Solving for T, $\ln(0.95) = \dfrac{-0.1499}{T}$ or $T = 2.83$ K.

(d) $\overline{E} = [n(2p\varepsilon)][2p\varepsilon] + \dfrac{[n(0)][0]}{n(2p\varepsilon) + n(0)} = \dfrac{[n(2p\varepsilon)/n(0)](2p\varepsilon)}{[n(2p\varepsilon) + n(0)] + 1} = \dfrac{\left[2e^{-2p\varepsilon/k_B T}\right][2p\varepsilon]}{2e^{-2p\varepsilon/k_B T} + 1}$

$= \dfrac{2p\varepsilon}{1 + (1/2)e^{2p\varepsilon/k_B T}}$.

As $T \to 0$, $\overline{E} \to 0$ and as $T \to \infty$, $E \to \dfrac{2p\varepsilon}{3/2} = \dfrac{4p\varepsilon}{3}$.

(e) $dE_{total} = \overline{N}E = \dfrac{2p\varepsilon N}{1 + (1/2)e^{2p\varepsilon/k_B T}}$

$C = \dfrac{dE_{total}}{dT} = \dfrac{(Nk_B/2)(2p\varepsilon/k_B T)^2 \, e^{2p\varepsilon/k_B T}}{\left[1 + (1/2)e^{2p\varepsilon/k_B T}\right]^2}$

(f)   By expanding e^x where $x = \dfrac{2p\varepsilon}{k_B T}$ one can show that $C \to 0$ for $T \to \infty$ as

$C = \left(\dfrac{8N}{9}\right)\left(\dfrac{p^2\varepsilon^2}{k_B^2}\right)\left(\dfrac{1}{T^2}\right)$ and $C \to 0$ for $T \to 0$ as $C = \dfrac{(2Nk_B)[2p\varepsilon/(k_B T)]^2}{e^{2p\varepsilon/k_B T}}$. To find the

maximum in $C = \left(\dfrac{Nk_B}{2}\right)(x^2)\left\{\dfrac{e^x}{\left[1 + (1/2)e^x\right]^2}\right\}$ set $\dfrac{dC}{dT} = 0$ or $\left(\dfrac{dC}{dx}\right)\left(\dfrac{dx}{dT}\right) = 0$. Taking

derivatives we get:

$$\left[\frac{-x^3 e^x}{\left(1 + (1/2)e^x\right)^2}\right]\left[\frac{2 + x - xe^x}{1 + (1/2)e^x}\right] = 0.$$

Setting the first factor equal to 0 yields the minima in C at $T = 0$ and $T = \infty$, while the second factor yields a maximum at the solution of the transcendental equation,

$\dfrac{x+2}{x-2} = \dfrac{e^x}{2}$. This transcendental equation has a solution at $x \approx 2.65$, which

corresponds to a temperature of $\dfrac{2p\varepsilon}{k_B T} = 2.65$ or $T = \dfrac{2p\varepsilon}{2.65 k_B} = \dfrac{0.1449}{2.65} = 0.0547$ K. The

expression for heat capacity can be rewritten as $C = A\left[e^x \dfrac{x}{\left(1 + (1/2)e^x\right)^2}\right]$ where

$A = \dfrac{Nk_B}{2}$ and $x = \dfrac{2p\varepsilon}{k_B T}$. Below is the sketch of C as a function of $\dfrac{2p\varepsilon}{k_B T}$.

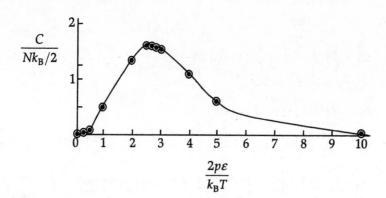

The heat capacity is the change of internal energy with temperature. For both large temperature $(T \to \infty)$ and low temperature $(T \to 0)$ the internal energy is constant and so the heat capacity is zero. At T approximately equal to 0.0547 K there is a rapid change of energy with temperature; so the heat capacity becomes large and reaches its maximum value.

10-9 $\bar{v} = \sqrt{\dfrac{8k_B T}{\pi m}}$. Using a molar weight of 55.85 g for iron gives the mass of an iron atom:

$m = \dfrac{55.85 \text{ g}}{6.02 \times 10^{23}} = 9.28 \times 10^{-26}$ kg. Thus, $\bar{v} = \sqrt{\dfrac{(8)(1.38 \times 10^{-23} \text{ J/K})(6\,000 \text{ K})}{(\pi)(9.28 \times 10^{-26} \text{ kg})}} = 1.51 \times 10^3$ m/s.

Since the speed of the emitting atoms is much less than c, we use the classical doppler shift, $f = f_0(1 \pm v/c)$. Then

$$\frac{\Delta f}{f_0} = \frac{f_{max}^+ - f_{max}^-}{f_0} = \frac{f_0(1+v/c) - f_0(1-v/c)}{f_0} = \frac{2v}{c} = \frac{(2)(1.51 \times 10^3 \text{ m/s})}{3.00 \times 10^8 \text{ m/s}} = 1.01 \times 10^{-5}$$

or 1 part per 100 000.

10-11

$$12E$$
$$11E$$
$$10E$$
$$9E$$
$$8E$$
$$7E$$
$$6E$$
$$5E$$

(2)   (3)   (4)   (5)   (6)   (7)   (8)   (9)

Thus,

$$\overline{n}_{0E} = \frac{1}{9}\times 2 + \frac{1}{9}\times 2 + \frac{1}{9}\times 2 + \frac{1}{9}\times 2 + \frac{1}{9}\times 2 + \frac{1}{9}\times 2 + \frac{1}{9}\times 2 + \frac{1}{9}\times 2 + \frac{1}{9}\times 2 = 2.00$$

$\overline{n}_{0E}$ through $\overline{n}_{5E} = 2.00$

$$\overline{n}_{6E} = 8\left(\frac{1}{9}\times 2\right) + \left(\frac{1}{9}\times 1\right) = 1.89$$

$$\overline{n}_{7E} = 7\left(\frac{1}{9}\times 2\right) + \left(\frac{1}{9}\times 1\right) + \left(\frac{1}{9}\times 1\right) = 1.78$$

$$\overline{n}_{8E} = 6\left(\frac{1}{9}\times 2\right) + \left(\frac{1}{9}\times 1\right) + \left(\frac{1}{9}\times 1\right) = 1.55$$

$$\overline{n}_{9E} = 4\left(\frac{1}{9}\times 2\right) + \left(\frac{1}{9}\times 1\right) + \left(\frac{1}{9}\times 1\right) + \left(\frac{1}{9}\times 1\right) = 1.22$$

$$\overline{n}_{10E} = \left(\frac{1}{9}\times 1\right) + \left(\frac{1}{9}\times 1\right) + \left(\frac{1}{9}\times 1\right) + \left(\frac{1}{9}\times 2\right) + \left(\frac{1}{9}\times 2\right) = 0.777$$

$$\overline{n}_{11E} = \left(\frac{1}{9}\times 2\right) + \left(\frac{1}{9}\times 1\right) + \left(\frac{1}{9}\times 1\right) = 0.444$$

$$\overline{n}_{12E} = \left(\frac{1}{9}\times 1\right) + \left(\frac{1}{9}\times 1\right) = 0.222$$

$$\overline{n}_{13E} = \left(\frac{1}{9}\times 1\right) = 0.111$$

$$\overline{n}_{14E} = 0.00$$

Minimum energy occurs for all levels filled up to $9E$, corresponding to a total energy of $90E$. So $E_F(0\ \text{K}) = 9E$. Using Equation 10.2 the following plot is obtained.

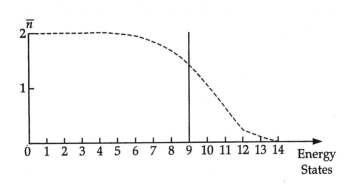

10-13   (a) $C = (3R)\left(\dfrac{\hbar\omega}{k_B T_E}\right)^2 \dfrac{e^{\hbar\omega/k_B T_E}}{\left(e^{\hbar\omega/k_B T_E} - 1\right)^2}$. For $T = T_E$, $k_B T_E = \hbar\omega$, so

$$C = (3R)\left(\frac{\hbar\omega}{\hbar\omega}\right)^2 \frac{e^{\hbar\omega/\hbar\omega}}{\left(e^{\hbar\omega/\hbar\omega} - 1\right)^2} = (3R)\frac{e}{(e-1)^2} = (3R)(0.920\,7) = 2.76R\,.$$

Using $R = 1.986$ cal/mol K $\Rightarrow C = 5.48$ cal/mol K.

(b)   From Figure 10.9, T_E lead ≈ 100 K, T_E aluminum ≈ 300 K, T_E silicon ≈ 500 K.

(c)   Using $C = (3R)\left(\dfrac{T_E}{T}\right)^2 \dfrac{e^{T_E/T}}{\left(e^{T_E/T}-1\right)^2} = (5.97 \ \text{cal/mol K})\left(\dfrac{T_E/T}{e^{T_E/T}-1}\right)^2 e^{T_E/T}$ heat capacities for

lead, aluminum, and silicon were obtained. These results can be summarized in the following tables.

Lead	$T_E = 100$ K		
T(K)	C(cal/(mol K))	T(K)	C(cal/(mol K))
50	4.32	250	5.92
100	5.49	300	5.94
150	5.74	350	5.96
200	5.83	400	6.09

Aluminum	$T_E = 300$ K		
T(K)	C(cal/(mol K))	T(K)	C(cal/(mol K))
50	0.535	250	5.30
100	2.96	300	5.509
150	4.32	350	5.62
200	4.97	400	5.70

Silicon	$T_E = 500$ K		
T(K)	C(cal/(mol K))	T(K)	C(cal/(mol K))
50	0.027	600	5.64
100	1.02	650	5.67
150	2.55	700	5.74
200	3.64	750	5.75
250	4.97	800	5.78
300	4.76	850	5.81
350	5.05	900	5.84
400	5.25	950	5.85
450	5.41	1 000	5.83
500	5.50	1 050	5.85
550	5.59	1 100	5.95

These values are now plotted on Figure 10.9 as shown.

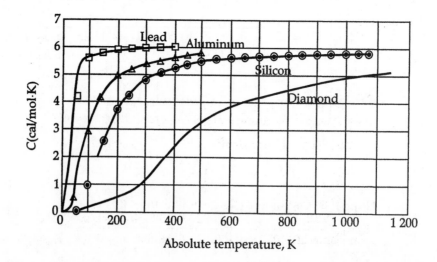

10-15   Al: $E_F = 11.63$ eV

(a) $E_F = \dfrac{h^2}{2m_e}\left(\dfrac{3n}{8\pi}\right)^{2/3}$ or $n = \dfrac{8\pi}{3}\left(\dfrac{2m_e E_F}{h^2}\right)^{3/2}$ so

$$n = \frac{8\pi}{3}\left[\frac{(2)(9.11\times10^{-31}\text{ kg})(11.63\text{ eV})(1.6\times10^{-19}\text{ J/eV})}{(6.625\times10^{-34}\text{ Js})^2}\right]^{3/2} = 1.80\times10^{29}\text{ free electrons/m}^3.$$

(b) $n' = \dfrac{\rho N_A}{M} = \dfrac{(2.7\text{ g/cm}^3)(6.02\times10^{23}\text{ atoms/mole})}{27\text{ g/mole}}$

$n' = 6.02\times10^{22}\text{ atoms/cm}^3 = 6.02\times10^{28}\text{ atoms/m}^3$

Valence $= \dfrac{n}{n'} = \dfrac{18\times10^{28}}{6\times10^{28}} = 3$

10-17   Equation 10.46 gives $E_F(0)$ in terms of $\dfrac{N}{V}$ as $E_F = \left(\dfrac{h^2}{2m}\right)\left(\dfrac{3N}{8\pi V}\right)^{2/3}$. Substituting the mass of a

proton, and noting that $A = 64$ for Zn, $m = 1.67\times10^{-27}$ kg; $N = \dfrac{A}{2} = 32$ and

$V = \dfrac{4}{3}\pi R^3 = \dfrac{4}{3}(\pi)(4.8\times10^{-15}\text{ m})^3 = 4.6\times10^{-43}\text{ m}^3$ yields

$$E_F = \frac{(6.62\times10^{-34})^2\text{ J}^2\text{s}^2}{3.34\times10^{-27}\text{ kg}} \times \left(\frac{(3)(32)}{(8\pi)(4.6\times10^{-43}\text{ m}^3)}\right)^{2/3} = 5.3\times10^{-12}\text{ J} = 33.4\text{ MeV}$$

$E_{av} = \dfrac{3}{5}E_F = 20$ MeV

These energies are of the correct order of magnitude for nuclear particles.

10-19 $f_{FD} = \left[e^{(E-E_F)/k_B T}+1\right]^{-1}$; $E_F = 7.05$ eV; $k_B T = (1.38\times10^{-23}\text{ J/K})(300\text{ K}) = 4.14\times10^{-21}\text{ J} = 0.025\,9$ eV

At $E = 0.99E_F$, $f_{FD} = \left[e^{-0.01E_F/k_B T}+1\right]^{-1} = \dfrac{1}{e^{-0.070\,5/0.025\,9}+1} = \dfrac{1}{1.065\,70} = 0.938$, thus 93.8%

probability.

10-21 $\rho = 0.971$ g/cm^3, $M = 23.0$ g/mole (sodium)

(a) $n = \dfrac{N_A \rho}{M}$

$n = (6.02\times10^{23}\text{ electrons/mole})(0.971\text{ g/cm}^3)(23.0\text{ g/mole})$

$n = 2.54\times10^{22}\text{ electrons/cm}^3 = 2.54\times10^{28}\text{ electrons/m}^3$

(b) $E_F = \dfrac{h^2}{2m}\left(\dfrac{3n}{8\pi}\right)^{2/3}$

$$E_F = \left[\frac{(6.625\times10^{-34}\text{ Js})^2}{(2\times9.11\times10^{-31}\text{ kg})}\right]\left[\frac{3\times2.54\times10^{28}\text{ electrons/m}^3}{8\pi}\right]^{2/3}$$

$E_F = 5.04\times10^{-19}\text{ J} = 3.15$ eV

(c) $v_F = \left(\dfrac{2E_F}{m}\right)^{1/2} = \left[\dfrac{2 \times 5.04 \times 10^{-19} \text{ J}}{9.11 \times 10^{-31} \text{ kg}}\right]^{1/2}$

$v_F = 1.05 \times 10^6$ m/s

10-23 $d = 1 \text{ mm} = 10^{-3}$ m; $V = \left(10^{-3} \text{ m}\right)^3 = 10^{-9}$ m^3

The density of states $= g(E) = CE^{1/2} = \left\{\dfrac{8(2)^{1/2}\pi m_e^{3/2}}{h^3}\right\}E^{1/2}$

$g(E) = 8(2)^{1/2}\pi\left(9.11 \times 10^{-31} \text{ kg}\right)^{3/2}\dfrac{\left[(4.0 \text{ eV})(1.6 \times 10^{-19} \text{ J/eV})\right]^{1/2}}{\left(6.626 \times 10^{-34} \text{ Js}\right)^3}$

$g(E) = \left(8.50 \times 10^{46}\right)\text{m}^{-3}\text{ J}^{-1} = \left(1.36 \times 10^{28}\right)\text{m}^{-3}\text{ eV}^{-1}$

$f_{FD}(E) = \dfrac{1}{e^{(E-E_F)/k_BT}+1}$ or

$f_{FD}(4.0 \text{ eV}) = \dfrac{1}{e^{(4.0-5.5)/(8.6\times10^{-5}\text{ eV/K})(300\text{ K})}+1} = \dfrac{1}{e^{-59}+1} = 1$

So the total number of electrons $= N = g(E)(\Delta E)Vf_{FD}(E)$ or
$N = \left(1.36 \times 10^{28} \text{ m}^{-3} \text{ eV}^{-1}\right)(0.025 \text{ eV})\left(10^{-9} \text{ m}^3\right)(1) = 3.40 \times 10^{17}$.

10-25   Use the equation $n(v) = \dfrac{4\pi N}{V}\dfrac{m}{(2\pi k_BT)^{3/2}}v^2 e^{-mv^2/(2k_BT)}$ where m is the mass of the O_2 molecule

in kg and $\dfrac{N}{V}$ is 10^4 molecules per cm^3. Rewrite the equation in the form

$n(v) = A_1\left(\dfrac{A_2}{T}\right)^{3/2}v^2 e^{-A_3 v^2}$ where $A_1 = \dfrac{4\pi N}{V}$, $A_2 = \dfrac{m}{2\pi k_B}$, and $A_3 = \dfrac{m}{2k_B}$. Use the exponential

format for large and small numbers to avoid computer errors.

(a)   For $T = 300$ K the equation can be rewritten as $n(v) = B_1 v^2 e^{-B_2 v^2}$ where

$B_1 = A_1\left(\dfrac{A_2}{300}\right)^{3/2}$ and $B_2 = \dfrac{A_3}{300}$. Do a 21 step loop for v from 0 to 2 000 m/s storing
$n300(i)$ as an array where $i = 1$ to 21 and corresponds to $v = 0$ to 2 000.

(b)   Repeat the calculation in (a) except that the A's are now divided by 1 000 and call the
array $n1000(j)$ where $j = 1$ to 21 and corresponds to $v = 0$ to 2 000.

(c)   Use a plot routine to obtain a graph similar to Figure 10.4 for the arrays obtained in
parts (a) and (b). To obtain the number of molecules with speeds between 800 m/s
and 1 000 m/s do a summation. The number of molecules
$= [n1 000(9)](100) + [n1 000(10)](100)$ where $n1 000(9)$ and $n1 000(10)$ is the number
calculated in (b) for speed 800 m/s and 900 m/s, respectively.

(d) $v_{rms} = \left(\dfrac{3k_BT}{m}\right)^{1/2}$; $v_{av} = \left(\dfrac{8k_BT}{\pi m}\right)^{1/2}$; $v_{mp} = \left(\dfrac{2k_BT}{m}\right)^{1/2}$. These quantities should appear

on your graph as shown in Figure 10.4.

10-27 (a)   For a metal $g(E) = \left[\dfrac{8(2)^{1/2} \pi m_e^{3/2}}{h^3} \right] E^{1/2} = DE^{1/2}$ where $D = \dfrac{8(2)^{1/2} \pi m_e^{3/2}}{h^3}$ and

$m_e = 0.511$ MeV$/c^2$ and $h = 4.136 \times 10^{-15}$ eV s. Using a loop calculate the array $g(E)$ for values of energy ranging from zero to 10 eV in steps of 0.5 eV. The array will be 21 dimensional, which can be plotted using a plot routine.

(b) $E_F(0) = \dfrac{h^2}{2m_e} \left(\dfrac{3N}{8\pi V} \right)^{2/3} = 7.05$ eV from Table 10.1. For $T = 0$ and $E_F < E$

$$f_{FD} = \frac{1}{e^{(E-E_F)/k_B T} + 1} = \frac{1}{e^\infty + 1} = 0$$
$$n(E) = 0.$$

For $T = 0$ and $E_F = E$, $n(E) = \left(\dfrac{D}{2} \right) E_F^{1/2}$. For $T = 0$ and $0 < E < E_F$ one has

$f_{FD} = \dfrac{1}{e^{-\infty} + 1} = 1$. Therefore $n(E) = g(E)$ where $g(E)$ is obtained from the array calculated in part (a). Use the same 0.5 eV steps in your loop.

(c) $n(E) = g(E) f_{FD}(E)$

Now calculate $f_{FD} = \dfrac{1}{e^{(E-E_F)/k_B T} + 1}$ where $T = 1\,000$ K in intervals of 0.5 eV for

$E = 0$ eV to 10 eV. E_F is determined for any temperature T numerically using the electron concentration

$$\frac{N}{V} = \int_0^\infty n(E) dE = D \int \frac{E^{1/2} dE}{e^{-(E-E_F)/k_B T} + 1} k_B T$$

that is of the order of 10^{-20}. The dependence of E_F on temperature is weak for metals and will not differ much from its value at 0 K up to several thousand kelvin and $E - E_F$ should be less than 10, which means $\dfrac{E - E_F}{k_B T}$ is large. Thus

$\dfrac{N}{V} \simeq D \int_0^\infty E^{1/2} e^{(E-E_F)k_B T} dE$. This can now be evaluated numerically. Once E_F is

determined then the Fermi Dirac distribution function, $f_{FD} = \dfrac{1}{e^{(E-E_F)/k_B T} + 1}$, can be

evaluated as an array using the same energy increments as before. The particle distribution function, $n(E)$, is the product of the arrays $g(E)$ and $f_{FD}(E)$. Now $n(E)$ can be plotted as a function of energy.

11

Molecular Structure

11-1 **(a)** We add the reactions $K + 4.34 \text{ eV} \rightarrow K^+ + e^-$ and $I + e^- \rightarrow I^- + 3.06 \text{ eV}$ to obtain $K + I \rightarrow K^+ + I^- + (4.34 - 3.06) \text{ eV}$. The activation energy is 1.28 eV.

(b) $$\frac{dU}{dr} = \frac{4\epsilon}{\sigma}\left[-12\left(\frac{\sigma}{r}\right)^{13} + 6\left(\frac{\sigma}{r}\right)^{7}\right]$$

At $r = r_0$ we have $\frac{dU}{dr} = 0$. Here $\left(\frac{\sigma}{r_0}\right)^{13} = \frac{1}{2}\left(\frac{\sigma}{r_0}\right)^{7}$, $\frac{\sigma}{r_0} = 2^{-1/6}$,

$$\sigma = 2^{-1/6}(0.305) \text{ nm} = \boxed{0.272 \text{ nm} = \sigma}.$$

Then also

$$U(r_0) = 4\epsilon\left[\left(\frac{2^{-1/6}r_0}{r_0}\right)^{12} - \left(\frac{2^{-1/6}r_0}{r_0}\right)^{6}\right] + E_a = 4\epsilon\left[\frac{1}{4} - \frac{1}{2}\right] + E_a = -\epsilon + E_a$$

$$\epsilon = E_a - U(r_0) = 1.28 \text{ eV} + 3.37 \text{ eV} = \boxed{4.65 \text{ eV} = \epsilon}.$$

(c) $$F(r) = -\frac{dU}{dr} = \frac{4\epsilon}{\sigma}\left[12\left(\frac{\sigma}{r}\right)^{13} - 6\left(\frac{\sigma}{r}\right)^{7}\right]$$

To find the maximum force we calculate $\frac{dF}{dr} = \frac{4\epsilon}{\sigma^2}\left[-156\left(\frac{\sigma}{r}\right)^{14} + 42\left(\frac{\sigma}{r}\right)^{8}\right] = 0$ when

$$\frac{\sigma}{r_{\text{rupture}}} = \left(\frac{42}{156}\right)^{1/6}$$

$$F_{\text{max}} = \frac{4(4.65 \text{ eV})}{0.272 \text{ nm}}\left[12\left(\frac{42}{156}\right)^{13/6} - 6\left(\frac{42}{156}\right)^{7/6}\right] = -41.0 \text{ eV/nm}$$

$$= -41.0\frac{1.6 \times 10^{-19} \text{ Nm}}{10^{-9} \text{ m}} = -6.55 \text{ nN}$$

Therefore the applied force required to rupture the molecule is $\boxed{+6.55 \text{ nN}}$ away from the center.

11-3   For the $l=1$ to $l=2$ transition, $\Delta E = hf = \dfrac{[2(2+1)-1(1+1)]\hbar^2}{2I}$ or $hf = \dfrac{2\hbar^2}{I}$. Solving for I gives

$$I = \frac{2\hbar^2}{hf} = \frac{h}{2\pi^2 f} = \frac{6.626\times10^{-34}\ \text{J s}}{(2\pi^2)(2.30\times10^{11}\ \text{Hz})} = 1.46\times10^{-46}\ \text{kg}\cdot\text{m}^2;\ \mu = \frac{m_1 m_2}{m_1 + m_2} = 1.14\times10^{-26}\ \text{kg},$$

$$R_0 = \left(\frac{I}{\mu}\right)^{1/2} = 0.113\ \text{nm, same as Example 11.1.}$$

11-5   (a)   The separation between two adjacent rotationally levels is given by $\Delta E = \left(\dfrac{\hbar^2}{I}\right)l$, where l is the quantum number of the higher level. Therefore

$$\Delta E_{10} = \frac{\Delta E_{65}}{6}$$

$$\lambda_{10} = 6\lambda_{65} = 6(1.35\ \text{cm}) = 8.10\ \text{cm}$$

$$f_{10} = \frac{c}{\lambda_{10}} = \frac{3.00\times10^{10}\ \text{cm/s}}{8.10\ \text{cm}} = 3.70\ \text{GHz}$$

(b) $\Delta E_{10} = hf_{10} = \dfrac{\hbar^2}{I};$

$$I = \frac{\hbar}{2\pi f_{10}} = \frac{1.055\times10^{-34}\ \text{J}\cdot\text{s}}{(2\pi)(3.70\times10^9\ \text{Hz})}$$

$$I = 4.53\times10^{-45}\ \text{kg}\cdot\text{m}^2$$

11-7   HCl molecule in the $l=1$ rotational energy level: $R_0 = 1.275$ Å, $E_{\text{rot}} = \left(\dfrac{\hbar^2}{2I}\right)l(l+1)$. For $l=1$,

$$E_{\text{rot}} = \frac{\hbar^2}{I} = \frac{I\omega^2}{2},\ \omega = \left(\frac{2\hbar^2}{I^2}\right)^{1/2} = \left(\frac{\hbar}{I}\right)\sqrt{2}$$

$$I = \left[\frac{m_1 m_2}{m_1 + m_2}\right]R_0^2 = \left[\frac{(1\ \text{u})(35\ \text{u})}{1\ \text{u} + 35\ \text{u}}\right]R_0^2 = \left[0.972\ 2\ \text{u}\times1.66\times10^{-27}\ \text{kg/u}\right]\times\left(1.275\times10^{-10}\ \text{m}\right)^2$$

$$= 2.62\times10^{-47}\ \text{kg}\cdot\text{m}^2$$

Therefore, $\omega = \left(\dfrac{\hbar}{I}\right)\sqrt{2} = \left[\dfrac{1.055\times10^{-34}\ \text{J}\cdot\text{s}}{2.62\times10^{-47}\ \text{kg}\cdot\text{m}^2}\right]\sqrt{2} = 5.69\times10^{12}\ \text{rad/s}.$

11-9 $\mu = \dfrac{m_1 m_2}{m_1 + m_2} = \dfrac{(1\ \text{u})(35\ \text{u})}{(1\ \text{u} + 35\ \text{u})} = \left(\dfrac{35}{36}\right)\text{u} = 1.62\times10^{-27}\ \text{kg}$

(a) $I = \mu R_0^2 = (1.62 \times 10^{-27} \text{ kg})(1.28 \times 10^{-10} \text{ m})^2 = 2.65 \times 10^{-47} \text{ kg} \cdot \text{m}^2$

$$E_{\text{rot}} = \left(\frac{\hbar}{2I}\right) l(l+1)$$

$$\frac{\hbar^2}{2I} = \frac{(1.054 \times 10^{-34} \text{ J} \cdot \text{s})^2}{2 \times 2.65 \times 10^{-47} \text{ kg} \cdot \text{m}^2} = 2.1 \times 10^{-22} \text{ J} = 1.31 \times 10^{-3} \text{ eV}$$

$$E_{\text{rot}} = (1.31 \times 10^{-3} \text{ eV}) l(l+1)$$

$l = 0$ $E_{\text{rot}} = 0$

$l = 1$ $E_{\text{rot}} = 2.62 \times 10^{-3} \text{ eV}$

$l = 2$ $E_{\text{rot}} = 7.86 \times 10^{-3} \text{ eV}$

$l = 3$ $E_{\text{rot}} = 1.57 \times 10^{-2} \text{ eV}$

(b) $U = \dfrac{Kx^2}{2}$, $U = 0.15 \text{ eV}$ when $x = 0.01 \text{ nm}$

$$(0.15 \text{ eV})(1.6 \times 10^{-19} \text{ J/eV}) = \frac{K(10^{-11} \text{ m})^2}{2}$$

$$K = 480 \text{ N/m}$$

$$f = \frac{1}{2\pi}\left(\frac{K}{\mu}\right)^{1/2} = \frac{1}{2\pi}\left[\frac{480}{1.62 \times 10^{-27}}\right]^{1/2} = 8.66 \times 10^{13} \text{ Hz}$$

(c) $E_{\text{vib}} = \left(v + \dfrac{1}{2}\right) hf$

$$hf = (6.63 \times 10^{-34} \text{ J s})(8.66 \times 10^{13} \text{ Hz}) = 5.74 \times 10^{-20} \text{ J} = 0.359 \text{ eV}$$

$$E_0 = \frac{hf}{2} = 2.87 \times 10^{-20} \text{ J} = 0.179 \text{ eV}$$

$$E = \frac{KA^2}{2}; \qquad 2.87 \times 10^{-20} \text{ J} = \frac{(480 \text{ N/m})A_0^2}{2}$$

$$A_0 = \left(\frac{2E}{K}\right)^{1/2} = 1.09 \times 10^{-11} \text{ m} = 0.109 \text{Å} = 0.010 \text{ 9 nm}$$

$$E_1 = \frac{3}{2} hf = 8.61 \times 10^{-20} \text{ J} = 0.538 \text{ eV}$$

$$A_1 = \left(\frac{2E}{K}\right)^{1/2} = 1.89 \times 10^{-11} \text{ m} = 0.189 \text{Å} = 0.018 \text{ 9 nm}$$

(d) $\dfrac{hc}{\lambda_{\text{max}}} = \Delta E_{\text{min}}$ or $\lambda_{\text{max}} = \dfrac{hc}{\Delta E_{\text{min}}}$

Rotational

$\Delta E_{\text{min}} = E_{l=1} - E_{l=0} = 2.62 \times 10^{-3} \text{ eV}$

$\quad hc = 12\,400 \text{ eV} \cdot \text{Å}$

$$\lambda_{\text{max}} = \frac{12\,400}{2.62 \times 10^{-3}} = 4.73 \times 10^6 \text{ Å} = 4.73 \times 10^{-4} \text{ m (microwave range).}$$

<u>Vibrational</u>
$$\Delta E_{min} = hf$$

$$\lambda_{max} = \frac{hc}{\Delta E_{min}} = \frac{hc}{hf} = \frac{c}{f} = \frac{3.00 \times 10^8 \text{ m/s}}{8.66 \times 10^{13} \text{ Hz}} = 3.46 \times 10^{-6} \text{ m} = 3.46 \text{ } \mu\text{m (infrared range)}.$$

11-11   The angular momentum of this system is $L = \dfrac{mvR_0}{2} + \dfrac{mvR_0}{2} = mvR_0$. According to Bohr theory,

L must be a multiple of $\hbar$, $L = mvR_0 = n\hbar$, or $v = \dfrac{n\hbar}{mR_0}$ with $n = 1, 2, \ldots$. The energy of rotation

is then

$$E_{rot} = \frac{1}{2}mv^2 + \frac{1}{2}mv^2 = m\left(\frac{n\hbar}{mR_0}\right)^2 = \frac{n^2\hbar^2}{mR_0^2}, \qquad n = 1, 2, \ldots .$$

From Equation 11.5 the allowed energies of rotation are

$$E_{rot} = \frac{\hbar^2}{2I_{cm}}\{l(l+1)\}, \qquad l = 0, 1, 2, \ldots$$

where I_{cm} is the moment of inertia about the center of mass. In the present case, we have

$$I_{cm} = m\left(\frac{R_0}{2}\right)^2 + m\left(\frac{R_0}{2}\right)^2 = \frac{mR_0^2}{2}.$$

Thus,

$$E_{rot} = \frac{\hbar^2}{mR_0^2}\{l(l+1)\} \qquad l = 0, 1, 2, \ldots .$$

We see that $l(l+1)$ replaces n^2 in the Bohr result. The two are indistinguishable for large quantum numbers (Correspondence Principle), but disagree markedly when n (or l) is small. In particular, E_{rot} can be zero according to Quantum Mechanics, while the minimum rotational energy in the Bohr theory is $\dfrac{\hbar^2}{mR_0^2}$ for $n = 1$.

11-13   At equilibrium separation R, U_{eff} is a minimum: $0 = \dfrac{dU_{eff}}{dr}\bigg|_{R_l} = \mu\omega_0^2(R_l - R_0) - \dfrac{l(l+1)\hbar^2}{\mu R^3}$ or

$R_l = R_0 + \dfrac{l(l+1)\hbar^2}{\mu^2\omega_0^2}\left(\dfrac{1}{R_l^3}\right)$. For $l \ll \dfrac{\mu\omega_0 R_0^2}{\hbar}$, the second term on the right represents a small

correction, and may be approximated by substituting for R its approximate value R_0 to get the

next approximation $R_l \approx R_0 + \dfrac{l(l+1)\hbar^2}{\mu^2\omega_0^2}\left(\dfrac{1}{R_0^3}\right)$. The value of U_{eff} at R_l is the energy offset U_0:

$$U_0 = U_{eff}(R) = \frac{1}{2}\mu\omega_0^2\left[\frac{l(l+1)\hbar^2}{\mu^2\omega_0^2 R_l^3}\right]^2 + \frac{l(l+1)\hbar^2}{2\mu R_l^2} = \left[\frac{l(l+1)\hbar^2}{2\mu R_l^2}\right]\left[\frac{l(l+1)\hbar^2}{\mu^2\omega_0^2 R_l^4} + 1\right]$$

$$\approx \frac{l(l+1)\hbar^2}{2\mu R_0^2}.$$

The curvature at the new equilibrium point is

$$\left.\frac{d^2U_{eff}}{dr^2}\right|_{R_l} = \mu\omega_0^2 + \frac{3l(l+1)\hbar^2}{\mu R_l^4}$$

and is identified with $\mu\omega_l^2$ to get the corrected oscillator frequency

$$\omega_l^2 = \omega_0^2 + \frac{3l(l+1)\hbar^2}{\mu^2 R_l^4} \approx \omega_0^2 + \frac{3l(l+1)\hbar^2}{\mu^2 R_0^4}$$

Since the second term on the right is small by assumption, ω_l differs little from ω_0, so that we may write $\omega_l^2 - \omega_0^2 = (\omega_l - \omega_0)(\omega_l + \omega_0) \approx 2\omega_0\Delta\omega$. The fractional change in frequency is then $\frac{\Delta\omega}{\omega_0} \approx \frac{3l(l+1)\hbar^2}{2\mu^2\omega_0^2 R_0^4}$.

11-15   The Morse levels are given by $E_{vib} = \left(v+\frac{1}{2}\right)\hbar\omega - \left(v+\frac{1}{2}\right)^2\frac{(\hbar\omega)^2}{4U_0}$. The excitation energy from level v to level $v+1$ is

$$\Delta E_{vib} = \left(v+\frac{3}{2}\right)\hbar\omega - \left(v+\frac{3}{2}\right)^2\frac{(\hbar\omega)^2}{4U_0} - \left(v+\frac{1}{2}\right)\hbar\omega + \left(v+\frac{1}{2}\right)^2\frac{(\hbar\omega)^2}{4U_0}$$

$$= \hbar\omega - \left\{\left(v+\frac{3}{2}\right)^2 - \left(v+\frac{1}{2}\right)^2\right\}\frac{(\hbar\omega)^2}{4U_0} = \hbar\omega\left[1 - (v+1)\left(\frac{\hbar\omega}{2U_0}\right)\right].$$

It is clear from this expression that ΔE_{vib} diminishes steadily as v increases. The excitation energy could never be negative, however, so that v must not exceed the value that makes ΔE_{vib} vanish: $1 = \frac{\hbar\omega}{2U_0}(v+1)$ or $v_{max} = \frac{2U_0}{\hbar\omega} - 1$. With this value for v, the vibrational energy is

$$E_{vib} = 2U_0 - \frac{1}{2}\hbar\omega - \frac{[2U_0 - (1/2)\hbar\omega]^2}{4U_0} = U_0 - \frac{(\hbar\omega)^2}{16U_0}.$$

If $\frac{2U_0}{\hbar\omega}$ is not an integer, then v_{max} and the corresponding E_{vib} will be somewhat smaller than the values given. However, the maximum vibrational energy will never exceed $U_0 - \frac{(\hbar\omega)^2}{16U_0}$.

11-19   To the left and right of the barrier site ψ is the waveform of a free particle with wavenumber $k = \left(\frac{2mE}{\hbar^2}\right)^{1/2}$:

$$\psi(x) = A\sin(kx) + B\cos(kx) \qquad 0 \leq x \leq \frac{L}{2}$$

$$\psi(x) = F\sin(kx) + G\cos(kx) \qquad \frac{L}{2} \leq x \leq L$$

The infinite walls at the edges of the well require $\psi(0) = \psi(L) = 0$, or $B = 0$ and $G = -F\tan(kL)$ leaving

$$\psi(x) = A\sin(kx) \qquad\qquad\qquad 0 \le x \le \frac{L}{2}$$

$$\psi(x) = F\{\sin(kx) - \tan(kL)\cos(kx)\} = C\sin(kx - kL) \quad \frac{L}{2} \le x \le L$$

For waves antisymmetric about the midpoint of the well, $\psi\left(\dfrac{L}{2}\right) = 0$ and the delta barrier is ineffective: the slope $\dfrac{d\psi}{dx}$ is continuous at $\dfrac{L}{2}$, leading to $C = +A$. For this case $\dfrac{kL}{2} = n\pi$, and

$$E_n = \frac{n^2\pi^2\hbar^2}{2m(L/2)^2} \qquad\qquad n = 1,\ 2,\ \dots$$

as befits an infinite well of width $\dfrac{L}{2}$.

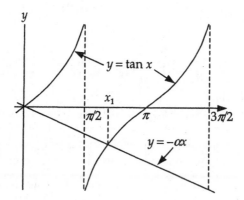

The remaining stationary states are waves symmetric about $\dfrac{L}{2}$, and require $C = -A$ for continuity of ψ. Their energies are found by applying the slope condition with $C = -A$ to get
$$-Ak\cos\left(\frac{kL}{2}\right) - Ak\cos\left(\frac{kL}{2}\right) = \left(\frac{2mS}{\hbar^2}\right)A\sin\left(\frac{kL}{2}\right) \text{ or } \tan\left(\frac{kL}{2}\right) = -\left(\frac{2\hbar^2}{mSL}\right)\left(\frac{kL}{2}\right).$$ Solutions to this equation may be found graphically as the intersections of the curve $y = \tan x$ with the line $y = -\alpha x$ having slope $-\alpha = -\dfrac{2\hbar^2}{mSL}$ (see the Figure above). From the points of intersection x_n we find $k_n = \dfrac{2x_n}{L}$ and $E_n = \dfrac{\hbar^2 k_n^2}{2m}$. Only values of x_n greater than zero need be considered, since the wave function is unchanged when k is replaced by $-k$, and $k = 0$ leads to $\psi(x) = 0$ everywhere. As $S \to \infty$ we see that $x_n \to n\pi$, giving $E_n = \dfrac{n^2\pi^2\hbar^2}{2m(L/2)^2}$ for $S \to \infty$ and $n = 1,\ 2,\ \dots$ the same energies found for the antisymmetric waves considered previously. Thus, in this limit the energy levels all are *doubly degenerate*. As $S \to 0$ the roots become
$$x_n = \frac{\pi}{2},\ \frac{3\pi}{2},\ \dots = \frac{n\pi}{2} \ (n \text{ odd}), \text{ giving } E_n = \frac{n^2\pi^2\hbar^2}{2mL^2} \quad n = 1,\ 3,\ \dots\ .$$ These are the energies for the symmetric waves of the infinite well with no barrier, as expected for $S = 0$.

The ground state wave is symmetric about $\frac{L}{2}$, and is described by the root x_1, which varies anywhere between $\frac{\pi}{2}$ and π according to S. The ground state energy is

$$E_1 = \frac{\hbar^2 (2x_1/L)^2}{2m} = \frac{2x_1^2 \hbar^2}{mL^2}.$$

The first excited state wave is antisymmetric, with energy

$$E_2 = \frac{\pi^2 \hbar^2}{2m(L/2)^2} = \frac{2\pi^2 \hbar^2}{mL^2},$$

which coincides with E_1 in the limit $S \to \infty$.

11-21   By trial and error, we discover that the choice $R = 1.44$ (bohr) minimizes the expression for E_{tot}, so that this is the equilibrium separation R_0.

The effective spring constant K is the curvature of $E_{\text{tot}}(R)$ evaluated at the equilibrium point $R_0 = 1.44$. Using the given approximation to the second derivative with an increment $\Delta R = 0.01$, we find

$$K = \left. \frac{d^2 E_{\text{tot}}}{dR^2} \right|_{R_0} \approx 1.03 .$$

(An increment ten times as large changes the result by less than one unit in the last decimal place.) This value for K is in $\left(\text{Ry}/\text{bohr}^2\right)$. The conversion to SI units is accomplished with the help of the relations $1\,\text{Ry} = 13.6\,\text{eV} = 2.176 \times 10^{-18}\,\text{J}$, and $1\,\text{bohr} = 0.529\,\text{Å} = 5.29 \times 10^{-11}\,\text{m}$. Then $K = 1.03\,\text{Ry}/\text{bohr}^2 = 801\,\text{J}/\text{m}^2 = 801\,\text{N/m}$. The result is larger than the experimental value because our neglect of electron-electron repulsion leads to a potential well much deeper than the actual one, producing a larger curvature.

12

The Solid State

12-1 $U_{Total} = U_{attractive} + U_{repulsive} = -\dfrac{\alpha k e^2}{r} + \dfrac{B}{r^m}$. At equilibrium, U_{Total} reaches its minimum value.

$\dfrac{dU_{Total}}{dr} = 0 = +\dfrac{\alpha k e^2}{r^2} - \dfrac{mB}{r^{m+1}}$. Calling the equilibrium separation r_0, we may solve for B

$$\dfrac{mB}{r_0^{m+1}} = \dfrac{\alpha k e^2}{r_0^2}$$

$$B = \dfrac{\alpha k e^2}{m r_0^{m-1}}.$$

Substituting into the expression for U_{Total} we find

$$U_0 = -\dfrac{\alpha k e^2}{r_0} + \dfrac{(\alpha k e^2/m) r_0^{m-1}}{r_0^m} = -\left(\dfrac{\alpha k e^2}{r_0}\right)\left(1 - \dfrac{1}{m}\right).$$

12-3 $U = -\alpha k \left(\dfrac{e^2}{r_0}\right)\left(1 - \dfrac{1}{m}\right)$

$U = -(1.747\,6)(9 \times 10^9 \ \text{Nm}^2/\text{C}^2)\left[\dfrac{(1.6 \times 10^{-19} \ \text{C})^2}{0.281 \times 10^{-9} \ \text{m}}\right]\left(1 - \dfrac{1}{8}\right)$

$U = -1.25 \times 10^{-18} \ \text{J} = -7.84 \ \text{eV}$. The ionic cohesive energy is $U = 7.84 \ \text{eV}/\text{Na}^+\text{-Cl}^-$ pair.

12-5 $U = -\dfrac{ke^2}{r} - \dfrac{ke^2}{r} + \dfrac{ke^2}{2r} + \dfrac{ke^2}{2r} - \dfrac{ke^2}{3r} - \dfrac{ke^2}{3r} + \dfrac{ke^2}{4r} + \dfrac{ke^2}{4r} - \ldots = -2k\left(\dfrac{e^2}{r}\right)\left[1 - \dfrac{1}{2} + \dfrac{1}{3} - \dfrac{1}{4} + \ldots\right]$

but $\ln(1 + x) = x - \dfrac{x^2}{2} + \dfrac{x^3}{3} - \dfrac{x^4}{4} + \ldots$ so $U = -\dfrac{(2\ln 2)ke^2}{r}$.

12-7   (a) $|U_0| = \left(\dfrac{\alpha k e^2}{r_0}\right)\left(1 - \dfrac{1}{m}\right) = \dfrac{(1.747\,6)(9.00 \times 10^9 \ \text{Nm}^2/\text{C}^2)(1.60 \times 10^{-19})^2}{0.314 \times 10^{-19}}$

$\left(1 - \dfrac{1}{9}\right) = 1.14 \times 10^{-18} \ \text{J} = 7.12 \ \text{eV}/\text{K}^+\text{-Cl}^-$

(b)   Atomic cohesive energy = ionic cohesive energy + energy needed to remove an electron from Cl^- – energy gained by adding the electron to $K^+ = 7.12 \ \text{eV} + 3.61 \ \text{eV} - 4.34 \ \text{eV} = 6.39 \ \text{eV}/\text{KCl}$.

12-9   (a) $$\int_0^\infty \left(\frac{N}{\tau}\right)e^{-t/\tau}dt = -Ne^{-t/\tau}\Big|_0^\infty = -N\left[e^{-\infty} - e^0\right] = N$$

(b) $$\bar{t} = \left(\frac{1}{N}\right)\int_0^\infty \left(\frac{tN}{\tau}\right)e^{-t/\tau}dt = \tau\int_0^\infty \left(\frac{t}{\tau}\right)e^{-t/\tau}\frac{dt}{\tau} = \tau\int_0^\infty ze^{-z}dz$$

$$z = u \qquad\qquad dv = e^{-z}dz$$
$$dz = du \qquad\qquad v = -e^{-z}$$

so $\int_0^\infty ze^{-z}dz = \left(-ze^{-z}\right)\Big|_0^\infty + \int_0^\infty e^{-z}dz = 0 - e^{-z}\Big|_0^\infty = 1$. Therefore, $\bar{t} = \tau$.

(c)   Similarly $\overline{t^2} = \left(\frac{1}{N}\right)\int_0^\infty \left(\frac{t^2 N}{\tau}\right)e^{-t/\tau}dt$. Integrating by parts twice, gives $\overline{t^2} = 2\tau^2$.

12-11   (a)   Equation 12.12 was $J = nev_d$. As $v_d = \mu E$, $J = ne\mu E$. Also comparing Equation 12.10, $v_d = \dfrac{e\tau E}{m_e}$, and $v_d = \mu E$, one has $\mu = \dfrac{e\tau}{m_e}$.

(b)   As $J = \sigma E$ and $J = J_{\text{electrons}} + J_{\text{holes}} = ne\mu_n E + pe\mu_p E$, $\sigma = ne\mu_n + pe\mu_p$

(c)   The electron drift velocity is given by

$$v_d = \mu_n E = \left(3\,900\ \text{cm}^2/\text{Vs}\right)\left(100\ \text{V/cm}\right) = 3.9\times10^5\ \text{cm/s}.$$

(d)   An intrinsic semiconductor has $n = p$. Thus

$$\sigma = ne\mu_n + pe\mu_p = pe\left(\mu_n + \mu_p\right) = \left(3.0\times10^{13}\ \text{cm}^{-3}\right)\left(1.6\times10^{-19}\ \text{C}\right)\left(5\,800\ \text{cm}^2/\text{Vs}\right)$$

$$= 0.028\ \text{A/V cm} = 0.028\ (\Omega\ \text{cm})^{-1} = 2.8(\Omega\ \text{m})^{-1}$$

$$\rho = \frac{1}{\sigma} = 0.36\ \Omega\ \text{m}$$

12-13   (a)   We assume all expressions still hold with v_{rms} replaced by v_F.

$$\tau = \frac{\sigma m_e}{ne^2}$$

$$\sigma = \frac{1}{\rho} = \left(1.60\times10^{-8}\right)^{-1}(\Omega\ \text{m})^{-1} = 6.25\times10^7\ (\Omega\ \text{m})^{-1}$$

$$n = \frac{\#\ \text{of}\ e^-}{m^3} = \left(\frac{1\,e^-}{\text{atom}}\right)\left(6.02\times10^{26}\ \text{atoms/k mole}\right)\left(10.5\times10^3\ \text{kg/m}^3\right)\left(\frac{1\ \text{kmole}}{108\ \text{g}}\right)$$

$$n = 5.85\times10^{28}\ e^-/\text{m}^3$$

so $\tau = \dfrac{\left(6.25\times10^7\right)(\Omega\ \text{m})^{-1}\left(9.11\times10^{-31}\ \text{kg}\right)}{\left(5.85\times10^{28}\ e^-/\text{m}^3\right)\left(1.6\times10^{-19}\ \text{C}\right)^2} = 3.80\times10^{-14}$ s (no change of course from Equation 12.10).

(b)   Now $L = v_F \tau$ and $v_F = \left(\dfrac{2E_F}{m}\right)^{1/2}$

$v_F = \left[\dfrac{2 \times 5.48 \text{ eV} \times 1.6 \times 10^{-19} \text{ J/eV}}{9.11 \times 10^{-31} \text{ kg}}\right]^{1/2} = 1.39 \times 10^6 \text{ m/s}.$

$L = (1.39 \times 10^6 \text{ m/s})(3.8 \times 10^{-14} \text{ s}) = 5.27 \times 10^{-8} \text{ m} = 527 \text{ Å} = 52.7 \text{ nm}$

(c)   The approximate lattice spacing in silver may be calculated from the density and the molar weight. The calculation is the same as the n calculation. Thus,
(# of Ag atoms)$/\text{m}^3 = 5.85 \times 10^{28}$. Assuming each silver atom fits in a cube of side, d,

$$d^3 = (5.85 \times 10^{28})^{-1} \text{ m}^3/\text{atom}$$

$$d = 2.57 \times 10^{-10} \text{ m}$$

So $\dfrac{L}{d} = \dfrac{5.27 \times 10^{-8}}{2.57 \times 10^{-10}} = 205.$

12-15   (a) $E_g = 1.14 \text{ eV}$ for Si

$hf = 1.14 \text{ eV} = (1.14 \text{ eV})(1.6 \times 10^{-19} \text{ J/eV}) = 1.82 \times 10^{-19} \text{ J}$

$f = 2.75 \times 10^{14} \text{ Hz}$

(b) $c = \lambda f; \ \lambda = \dfrac{c}{f} = \dfrac{3 \times 10^8 \text{ m/s}}{2.75 \times 10^{14} \text{ Hz}} = 1.09 \times 10^{-6} \text{ m}$

$\lambda = 1\,090 \text{ nm}$ (in the infrared region)

12-17   (a)   Potential

$\psi_I = Ae^{Kx}$ $K\hbar = [2m(U-E)]^{1/2}$

$\psi_{II} = B\cos kx + C\sin kx$ $k\hbar = (2mE)^{1/2}$

$\psi_{III} = De^{-Kx}$

In region I and III the wave equation has the form $\dfrac{d^2\psi(x)}{dx^2} = K^2\psi(x)$ with

$K = \dfrac{[2m(U-E)]^{1/2}}{\hbar}$. This equation has solutions of the form

$$\psi_I(x) = Ae^{Kx} \text{ for } x \le 0 \quad \text{(region I)}$$
$$\psi_{III} = De^{-Kx} \text{ for } x \ge 0 \quad \text{(region III)}$$

In region II where $U(x)=0$ we have $\dfrac{d^2\psi(x)}{dx^2}=-k^2\psi(x)$ with $k=\dfrac{[2mE]^{1/2}}{\hbar}$. This equation has trigonometric solutions

$$\psi_{II}(x)=B\cos kx+C\sin kx \qquad 0\leq x\leq a$$

with $k=\dfrac{(2mE)^{1/2}}{\hbar}$. The wave function and its slope are continuous everywhere, and in particular at the well edges $x=0$ and $x=a$. Thus, we must require

$$A=B \qquad\qquad [\text{continuity of } \psi(x) \text{ at } x=0]$$
$$KA=kC \qquad\qquad \left[\text{continuity of } \frac{d\psi(x)}{dx} \text{ at } x=0\right]$$
$$B\cos kx+C\sin kx \doteq De^{-Kx} \qquad [\text{continuity of } \psi(x) \text{ at } x=a]$$
$$-Bk\sin kx+Ck\cos kx=-DKe^{-Kx} \qquad \left[\text{continuity of } \frac{d\psi(x)}{dx} \text{ at } x=a\right]$$

There are four equations in the four coefficients A, B, C, D. Use the first equation to eliminate A. Then from the second equation we obtain $B=\left(\dfrac{k}{K}\right)C$. Divide the last two equations to eliminate D.

$$\frac{-Bk\sin kx+Ck\cos kx}{B\cos kx+C\sin kx}=-\frac{DKe^{-Kx}}{De^{-Kx}}.$$

Cross multiply, gather terms and write B in terms of C. Then we have

$$\left(-\frac{k^2}{K}\right)Ck\sin ka+Ck\cos ka=-K\left(\frac{k}{K}\right)C\cos ka-KC\sin ka.$$

Divide out C and gather terms to obtain $\left(K^2-k^2\right)\sin ka=-2ka\cos k$. Now substitute $k=(2mE)^{1/2}\left[\dfrac{2m(U-E)}{\hbar^2}\right]^{1/2}\cos ka$. This equation simplifies to:

$U\sin ka=-2[E(U-E)]^{1/2}\cos ka$, which is a transcendental equation for the bound energy states. Rearranging,

$$\tan^2 ka=\tan^2\left[\frac{(2mE)^{1/2}}{\hbar}\right]a=\frac{4E(U-E)}{U^2}.$$

(b)   The energy equation is a transcendental equation and can be solved for the roots, E_n by using Newton's root formula as an iterative method employing a computer. If you know the form of $f(x)$ then you can approximate the value of x for which $f(x)=0$. Choose an initial value of x. The energy equation can be written as

$$f(E)=\tan^2\left[\frac{(2mE)^{1/2}}{\hbar}\right]a-[4E(U-E)]U^2=0.$$

In this problem approximate by using the energy for an electron in a well. The first guess energy is: $E_n = \dfrac{n^2\pi^2\hbar^2}{2m_e a + 2\delta}$ where

$$\delta = \frac{1}{K} \approx \frac{197.3 \ \text{eV nm}/c}{2(0.511\times 10^6 \ \text{eV}/c^2)(100 \ \text{eV})} = 0.019 \ 3 \ \text{nm}$$

and so

$$E_n = \frac{n^2\pi^2(197.3 \ \text{eV nm}/c)}{2(0.511\times 10^6 \ \text{eV}/c^2)(0.10 \ \text{nm} + 0.039 \ \text{nm})^2} = n^2(19.5 \ \text{eV})$$

$$E_1 = 19.5 \ \text{eV}$$

$$E_2 = (2)^2(19.5 \ \text{eV}) = 78.0 \ \text{eV}$$

$$E_3 = (3)^2(19.5 \ \text{eV}) = 175.5 > U \text{ therefore unbound}$$

These values seem reasonable since there are only two bound states. The next step in Newton's method is to calculate $f(x)$ and $f'(x)$ at the first guess value. Then use the definition of slope and tangent:

$$x_1 - x_2 = \frac{f(x)}{f'(x)} \text{ or } x_2 = x_1 - \frac{f(x)}{f'(x)}.$$

Use x_2 as a new estimate to evaluate x_3, etc. Monitor x and $f(x)$ for convergence and divergence. Use the first term of two of the Taylor series for the first guess of $f'(x)$. Thus our first guess would be $x_1 = E = 19.5 \ \text{eV}$ and

$$f(E) = \tan^2 ka - \frac{4E(U-E)}{U^2}$$

$$f'(E) = \left(\frac{2a}{\hbar}\right)\left(\frac{2m_e}{E}\right)^{1/2} \frac{\sin\left[(a/\hbar)(2mE)^{1/2}\right]}{\cos^3\left[(a/\hbar)(2m_eE)^{1/2}\right]} + \left(\frac{4}{U^2}\right)(2E-U)$$

where $U = 100 \ \text{eV}$ and $E = 19.5 \ \text{eV}$. Calculate x_2 and keep repeating, watching for convergence.

(c)

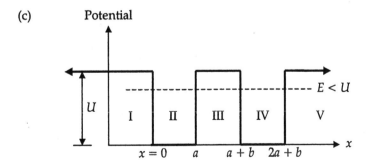

$$\psi_{\mathrm{I}} = Ae^{Kx} \qquad\qquad K\hbar = [2m(U-E)]^{1/2}$$
$$\psi_{\mathrm{II}} = B\cos kx + C\sin kx \qquad k\hbar = (2mE)^{1/2}$$
$$\psi_{\mathrm{III}} = De^{Kx} + E'e^{-Kx}$$
$$\psi_{\mathrm{IV}} = F\cos kx + G\sin kx$$
$$\psi_{\mathrm{V}} = He^{-Kx}$$

At $x=0$, $\psi_{\mathrm{I}} = \psi_{\mathrm{II}}$. Therefore $A=B$, $\dfrac{d\psi_{\mathrm{I}}}{dx} = \dfrac{d\psi_{\mathrm{II}}}{dx}$ yields $KA = kC$. Similarly at $x=a$:

$\psi_{\mathrm{II}} = \psi_{\mathrm{III}}$, $B\cos ka + C\sin ka = De^{Ka} + E'e^{-Ka}$ and $\dfrac{d\psi_{\mathrm{II}}}{dx} = \dfrac{d\psi_{\mathrm{III}}}{dx}$,

$-Bk\cos ka + Ck\cos ka = KDe^{Ka} - KE'e^{-Ka}$. Substitute $C = \dfrac{KA}{k}$ and $B=A$ to obtain

$$A\cos ka + \left(\frac{KA}{k}\right)\sin ka = De^{Ka} + E'e^{-Ka}$$
$$-Ak\sin ka + \left(\frac{KA}{k}\right)\cos ka = KDe^{Ka} - KE'e^{-Ka}$$

Solve for A in each equation and equate quantities to obtain

$$De^{Ka} + \frac{E'e^{-Ka}}{\cos ka + (K/k)\sin ka} = KDe^{Ka} - \frac{KE'e^{-Ka}}{-k\sin ka + K\cos ka}.$$

Clear denominators and gather terms. After some algebra one obtains

$$\frac{D}{E'} = \frac{-K(\cos ka + (K/a)\sin ka) - (-k\sin ka + K\cos ka)e^{-Ka}}{[-k\sin ka + K\cos ka - K(\cos ka + (K/k)\sin ka)]e^{Ka}}.$$

This can be simplified to obtain

$$\frac{D}{E'} = \frac{2e^{-2Ka}[\cos ka + (1/2)[(K/k) - k/K]\sin ka]}{[(k/K) + K/k]\sin ka}.$$

Impose the continuity conditions at $x = a+b$ and let $\alpha = k(a+b)$ and

$$\beta = K(a+b)$$
$$\psi_{\mathrm{III}} = \psi_{\mathrm{IV}}$$

$$De^{\beta} + E'e^{-\beta} = F\cos\alpha + G\sin\alpha \Rightarrow \frac{F\cos\alpha + G\sin\alpha}{De^{\beta} + E'e^{-\beta}} = 1 \text{, and } \frac{d\psi_{\mathrm{III}}}{dx} = \frac{d\psi_{\mathrm{IV}}}{dx}$$

$$KDe^{\beta} + KE'e^{-\beta} = -Fk\cos\alpha + Gk\sin\alpha \Rightarrow \frac{-Fk\cos\alpha + Gk\sin\alpha}{KDe^{\beta} + KE'e^{-\beta}} = 1.$$

Set quantities equal to 1 equal to each other and clear fractions to obtain

$$(F\cos\alpha + G\sin\alpha)(KDe^{\beta} + KE'e^{-\beta}) = (-Fk\cos\alpha + Gk\sin\alpha)(De^{\beta} + E'e^{-\beta}).$$

Divide by E' and gather terms to obtain

$$FK\left[\left(\frac{D}{E'}\right)\cos\alpha+\left(\frac{k}{K}\right)\left(\frac{D}{E'}\right)\sin\alpha\right]e^{\beta}-Fk\left[\cos k\alpha+\left(\frac{k}{K}\right)\sin\alpha\right]e^{-\beta}$$

$$=GK\left[\left(\frac{k}{K}\right)\left(\frac{D}{E'}\right)\cos\alpha-\left(\frac{D}{E'}\right)\sin\alpha\right]e^{\beta}+GK\left[\sin\alpha+\left(\frac{k}{K}\right)\left(\frac{D}{E'}\right)\cos\alpha\right]e^{-\beta}$$

Divide through by G and $\left(\dfrac{k}{K}\right)$ to obtain

$$\frac{F}{G}=\frac{(D/E')e^{\beta}[\cos\alpha-(K/k)\sin\alpha]+e^{-\beta}[\cos\alpha+(K/k)\sin\alpha]}{(D/E')e^{\beta}[\sin\alpha+(K/k)\cos\alpha]+e^{-\beta}[\sin\alpha-(K/k)\cos\alpha]}$$

at $x=2a$, $\psi_{\mathrm{IV}}=\psi_{\mathrm{V}}$, $F\cos k(2a+b)+G\sin k(2a+b)=He^{-Kx}$ and $\dfrac{d\psi_{\mathrm{IV}}}{dx}=\dfrac{d\psi_{\mathrm{V}}}{dx}$ and

dividing by $(-K)$ we obtain $\dfrac{1}{K}[Fk\sin k(2a+b)-G\cos k(2a+b)]=He^{-Kx}$. Both equations are equal to the same quantity so set equal to each other.

$$F\cos k(2a+b)+G\sin k(2a+b)=\frac{1}{K}[Fk\sin k(2a+b)-G\cos k(2a+b)].$$

Now gather terms and divide by G and $\left(-\dfrac{k}{K}\right)$ to obtain

$$\frac{F}{G}=\frac{\cos k(2a+b)+(K/k)\sin k(2a+b)}{\sin k(2a+b)-(K/k)\cos k(2a+b)}.$$

Equating the two expressions for $\dfrac{F}{G}$

$$\frac{\cos k(2a+b)+(K/k)\sin k(2a+b)}{\sin k(2a+b)-(K/k)\cos k(2a+b)}=\frac{(D/E')e^{\beta}[\cos\alpha-(K/k)\sin\alpha]+e^{-\beta}[\cos\alpha+(K/k)\sin\alpha]}{(D/E')e^{\beta}[\sin\alpha+(K/k)\cos\alpha]+e^{-\beta}[\sin\alpha-(K/k)\cos\alpha]}$$

Bringing all terms to one side gives a transcendental equation in E

$$f(E)=\frac{(D/E')e^{\beta}[\cos\alpha-(K/k)\sin\alpha]+e^{-\beta}[\cos\alpha+(K/k)\sin\alpha]}{(D/E')e^{\beta}[\sin\alpha+(K/k)\cos\alpha]+e^{-\beta}[\sin\alpha-(K/k)\cos\alpha]}$$

$$-\frac{\cos k(2a+b)+(K/k)\sin k(a+b)}{\sin k(2a+b)-(K/k)\cos k(2a+b)}=0$$

with U, a, and b as parameters. This equation can be solved numerically with Newton roots method used in the solution to 12-17(b). The form of the program will depend strongly on the computer language used, including its subroutine (function, module) structure. Assume you can write a module to calculate $f(E)$ where $a=b=1$ and $U=100$. Output tabular values of E and $f(E)$ and/or graph E and $f(E)$. The Newton method requires both function and its derivative to be used. This is algebraically complicated so that it proves more practical to use a more interactive program. Use the computer to calculate $f(E)$ for any E you enter. Use trial and error to converge to the values of E for which $f(E)$ changes sign. Those are the values of E, which satisfy the equation and are the bound states of the double square well.

The search procedure is: Guess one value of E and calculate $f(E)$. Guess a second value of E, not very different and calculate $f(E)$. If the sign of $f(E)$ changes, interpolate a new E and calculate its $f(E)$. If the sign of $f(E)$ did not change, extrapolate in a direction toward the smaller $|f(E)|$. Continue until ΔE, which causes $f(E)$ to change spin, is small enough for your needs. That is, less than 1 eV for this problem, since you are looking for other splittings of the single-well energies at 19 eV and 70 eV.

12-19 (a)
$$\frac{dm}{d\lambda} = \frac{d}{d\lambda}\left\{\frac{2Ln}{\lambda}\right\} = \left(\frac{2L}{\lambda}\right)\left(\frac{dn}{d\lambda} - \frac{n}{\lambda}\right)$$
Replacing dm and $d\lambda$ with Δm and $\Delta\lambda$ yields

$$\Delta\lambda = \frac{\lambda^2 \Delta m}{2L}\left(\frac{\lambda dn}{d\lambda} - n\right)$$

or $|\Delta\lambda| = \frac{\lambda^2}{2L}\left(n - \frac{\lambda dn}{d\lambda}\right)$. Since $\Delta\lambda$ is negative for $\Delta m = +1$.

(b)
$$|\Delta\lambda| = \frac{(837\times10^{-9}\text{ m})^2}{(0.6\times10^{-3}\text{ m})}\left[3.58 - (837\text{ nm})(3.8\times10^{-4}\text{ nm}^{-1})\right] = 3.6\times10^{-10}\text{ m} = 0.38\text{ nm}$$

(c)
$$|\Delta\lambda| = \frac{(633\times10^{-9}\text{ m})^2}{(0.6\times10^0\text{ m})(1)} = 6.7\times10^{-13}\text{ m} = 0.000\,67\text{ nm} = 6.7\times10^{-4}\text{ nm}$$
The controlling factor is <u>cavity length</u>, L.

12-21 (a)   ┌─────────────────────┐
              │ See the figure below. │
              └─────────────────────┘

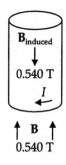

B_{induced}
$\downarrow$
0.540 T

I
$\leftarrow$

$\uparrow$ B $\uparrow$
0.540 T

(b)   For a surface current around the outside of the cylinder as shown,

$$B = \frac{N\mu_0 I}{\ell} \quad\text{or}\quad NI = \frac{B\ell}{\mu_0} = \frac{(0.540\text{ T})(2.50\times10^{-2}\text{ m})}{(4\pi\times10^{-7})\text{ T}\cdot\text{m/A}} = \boxed{10.7\text{ kA}}.$$

12-23 (a) $\Delta V = IR$
              If $R = 0$, then $\Delta V = 0$, even when $I \neq 0$.

(b)   The graph shows a direct proportionality.

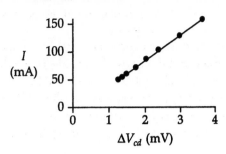

$$\text{Slope} = \frac{1}{R} = \frac{\Delta I}{\Delta V} = \frac{(155 - 57.8)\ \text{mA}}{(3.61 - 1.356)\ \text{mV}} = 43.1\ \Omega^{-1}$$

$$R = \boxed{0.023\ 2\ \Omega}$$

(c)   Expulsion of magnetic flux and therefore fewer current-carrying paths could explain the decrease in current.

12-25 (a)   The currents to be plotted are

$$I_D = \left(10^{-6}\ \text{A}\right)\left(e^{\Delta V/0.025\ \text{V}} - 1\right),\ I_W = \frac{2.42\ \text{V} - \Delta V}{745\ \Omega}$$

The two graphs intersect at $\Delta V = 0.200\ \text{V}$. The currents are then

$$I_D = \left(10^{-6}\ \text{A}\right)\left(e^{0.200\ \text{V}/0.025\ \text{V}} - 1\right) = 2.98\ \text{mA}$$

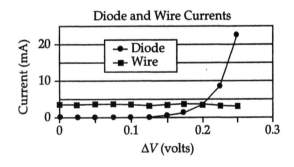

$$I_W = \frac{2.42\ \text{V} - 0.200\ \text{V}}{745\ \Omega} = 2.98\ \text{mA}. \text{ They agree to three digits.} \therefore I_D = I_W = \boxed{2.98\ \text{mA}}$$

(b) $\dfrac{\Delta V}{I_D} = \dfrac{0.200\ \text{V}}{2.98 \times 10^{-3}\ \text{A}} = \boxed{67.1\ \Omega}$

(c) $\dfrac{d(\Delta V)}{dI_D} = \left[\dfrac{dI_D}{d(\Delta V)}\right]^{-1} = \left[\dfrac{10^{-6}\ \text{A}}{0.025\ \text{V}}\, e^{0.200\ \text{V}/0.025\ \text{V}}\right]^{-1} = \boxed{8.39\ \Omega}$

13

Nuclear Structure

13-1 $R = R_0 A^{1/3}$ where $R_0 = 1.2$ fm;

(a) $A = 4$ so $R_{He} = (1.2)(4)^{1/3}$ fm $= 1.9$ fm

(b) $A = 238$ so $R_U = (1.2)(238)^{1/3}$ fm $= 7.44$ fm

(c) $\dfrac{R_U}{R_{He}} = \dfrac{7.44 \text{ fm}}{1.9 \text{ fm}} = 3.92$

13-3 $\dfrac{\rho_{NUC}}{\rho_{ATOMIC}} = \dfrac{M_{NUC}/V_{NUC}}{M_{ATOMIC}/V_{ATOMIC}}$ and approximately; $M_{NUC} = M_{ATOMIC}$. Therefore

$\dfrac{\rho_{NUC}}{\rho_{ATOMIC}} = \left(\dfrac{r_0}{R}\right)^3$ where $r_0 = 0.529$ Å $= 5.29 \times 10^{-11}$ m and $R = 1.2 \times 10^{-15}$ m (Equation 13.1

where $A = 1$). So that $\dfrac{\rho_{NUC}}{\rho_{ATOMIC}} = \left(\dfrac{5.29 \times 10^{-11} \text{ m}}{1.2 \times 10^{-15} \text{ m}}\right)^3 = 8.57 \times 10^{13}$.

13-5 (a) The initial kinetic energy of the alpha particle must equal the electrostatic potential energy of the two particle system at the distance of closest approach; $K_\alpha = U = \dfrac{kqQ}{r_{min}}$

and $r_{min} = \dfrac{kqQ}{K_\alpha} = \dfrac{(9 \times 10^9 \text{ N m}^2/\text{C}^2)2(79)(1.6 \times 10^{-19} \text{ C})^2}{\left[0.5 \text{ MeV}(1.6 \times 10^{-13} \text{ J/MeV})\right]} = 4.55 \times 10^{-13}$ m.

(b) Note that $K_\alpha = \dfrac{1}{2}mv^2 = \dfrac{kqQ}{r_{min}}$, so

$v = \left[\dfrac{2kqQ}{mr_{min}}\right]^{1/2} = \left[\dfrac{2(9 \times 10^9 \text{ N m}^2/\text{C}^2)2(79)(1.6 \times 10^{-19} \text{ C})^2}{4(1.67 \times 10^{-27} \text{ kg})(3 \times 10^{-13} \text{ m})}\right]^{1/2} = 6.03 \times 10^6$ m/s

13-7 $E = -\mu \cdot B$ so the energies are $E_1 = +\mu B$ and $E_2 = -\mu B$. $\mu = 2.792\,8\mu_n$ and $\mu_n = 5.05 \times 10^{-27}$ J/T
$\Delta E = 2\mu B = 2 \times 2.792\,8 \times 5.05 \times 10^{-27}$ J/T $\times 12.5$ T $= 3.53 \times 10^{-25}$ J $= 2.2 \times 10^{-6}$ eV

13-9   We need to use the procedure to calculate a "weighted average." Let the fractional abundances be represented by $f_{63} + f_{65} = 1$; then $\dfrac{f_{63}m\left(^{63}Cu\right) + f_{65}m\left(^{65}Cu\right)}{\left(f_{63} + f_{65}\right)} = m_{Cu}$. We find

$$f_{63} = \frac{m\left(^{65}Cu\right) - m_{Cu}}{m\left(^{65}Cu\right) - m\left(^{63}Cu\right)},\ f_{63} = \frac{64.95\ u - 63.55\ u}{64.95\ u - 62.95\ u} = 0.30\ \text{or}\ 30\%\ \text{and}\ f_{65} = 1 - f_{63} = 0.70\ \text{or}\ 70\%.$$

13-11 $\dfrac{E_b}{A} = \dfrac{1}{3}[1(1.007\ 276\ u) + 2(1.008\ 665\ u) - 3.016\ 05\ u](931.5\ MeV/u) = 2.657\ MeV/nucleon$

13-13   (a)   The neutron to proton ratio, $\dfrac{A - Z}{Z}$ is greatest for $^{139}_{55}Cs$ and is equal to 1.53.

(b)   Using $E_b = C_1 A - C_2 A^{2/3} - C_3(Z(Z-1))A^{-1/3} - C_4 \dfrac{(N-Z)^2}{A}$ the only variation will be in the coefficients of C_3 and C_4 since the isotopes have the same A number. For $^{139}_{59}Pr$

$$E_b = (15.7)(139) - (17.8)(139)^{2/3} - 0.71(59)(139)^{1/3} - \frac{23.6(21)^2}{139} = 1\,160.8\ MeV$$

$$\frac{E_b}{A} = \frac{E_b}{139} = 8.351\ MeV$$

For $^{139}_{57}La$

$$E_b = (15.7)(139) - (17.8)(139)^{2/3} - 0.71(55)(54)(139)^{1/3} - \frac{23.6(25)^2}{139} = 1\,161.1\ MeV$$

$$\frac{E_b}{A} = \frac{E_b}{139} = 8.353\ MeV$$

For $^{139}_{55}Cs$

$$E_b = (15.7)(139) - (17.8)(139)^{2/3} - 0.71(55)(54)(139)^{1/3} - \frac{23.6(29)^2}{139} = 1\,154.9\ MeV$$

$$\frac{E_b}{A} = \frac{E_b}{139} = 8.308\ MeV$$

^{139}La has the largest binding energy per nucleon of 8.353 MeV

(c)   The mass of the neutron is greater than the mass of a proton therefore expect the nucleus with the largest N and smallest Z to weigh the most: $^{139}_{55}Cs$ with a mass of 138.913 u.

13-15   Use Equation 13.4, $E_b = \left[ZM(H) + Nm_n - M\left(^A_Z X\right)\right]$

(a)   For $^{20}_{10}Ne$;

$$E_b = [10(1.007\ 825\ u) + 10(1.008\ 665) - (19.992\ 436\ u)](931.494\ MeV/u) = 160.650$$

$$\frac{E_b}{A} = 8.03\ MeV/nucleon$$

(b)   For $^{40}_{20}$Ca

$$E_b = [20(1.007\,825\ \text{u}) + 20(1.008\,665) - (39.962\,591\ \text{u})](931.494\ \text{MeV/u}) = 342.053$$

$$\frac{E_b}{A} = 8.55\ \text{MeV/nucleon}$$

(c)   For $^{93}_{41}$N$_b$;

$$E_b = [41(1.007\,825\ \text{u}) + 52(1.008\,665) - (92.906\,377\ \text{u})](931.494\ \text{MeV/u}) = 805.768$$

$$\frac{E_b}{A} = 8.66\ \text{MeV/nucleon}$$

(d)   For $^{197}_{79}$Au

$$E_b = [79(1.007\,825\ \text{u}) + 118(1.008\,665) - (196.966\,543\,1\ \text{u})](931.494\ \text{MeV/u}) = 1\,559.416$$

$$\frac{E_b}{A} = 7.92\ \text{MeV/nucleon}$$

13-17 $\Delta E = E_{bf} - E_{bi}$

For $A = 200$; $\dfrac{E_b}{A} = 7.8\ \text{MeV}$ so

$$E_{bi} = (A_i)(7.8\ \text{MeV}) = (200)(7.8) = 1\,560\ \text{MeV}$$

For $A \approx 100$; $\dfrac{E_b}{A} \approx 8.6\ \text{MeV}$ so

$$E_{bf} = (2)(100)(8.6\ \text{MeV}) = (200)(8.6) = 1\,720\ \text{MeV}$$
$$\Delta E = E_{bf} - E_{bi} = 1\,720\ \text{MeV} - 1\,560\ \text{MeV} = 160\ \text{MeV}$$

13-19   (a)   The potential at the surface of a sphere of charge q and radius r is $V = \dfrac{kq}{r}$. If a thin shell of charge dq (thickness dr) is <u>added</u> to the sphere, the increase in electrostatic potential energy will be $dU = Vdq = \left(\dfrac{kq}{r}\right)dq$. To build up a sphere with final radius R, the total energy will be $U = \int_0^R \left(\dfrac{kq}{r}\right)dq$; where $q = \dfrac{4}{3}\pi r^3 \rho = \dfrac{4}{3}\pi r^3 \left[\dfrac{Ze}{4\pi R^3/3}\right] = \left(\dfrac{Ze}{R^3}\right)r^3$ so that

$$dq = \left(\frac{3Ze}{R^3}\right)r^2 dr$$
$$U = \left(\frac{3kZ^2e^2}{R^6}\right)\int_0^R r^4 dr = \frac{3k(Ze)^2}{5R}$$

(b)   When $N = Z = \dfrac{A}{2}$, $R = R_0 A^{1/3}$ and $R_0 = 1.2 \times 10^{-15}$ m

$$U = \frac{3k(Ze)^2}{5R} = \frac{(3/5)(8.988 \times 10^9 \text{ N m}^2/\text{C}^2)(A/2)^2(1.602 \times 10^{-19} \text{ C})^2}{(1.2 \times 10^{-15} \text{ m})A^{1/3}}$$

$$= 2.88 \times 10^{-14} \left(A^{5/3}\right) \text{J}$$

(c)   For $A = 30$, $U = 8.3 \times 10^{-12}$ J $= 52.1$ MeV.

13-21   (a)   Write Equation 13.10 as $\dfrac{R}{R_0} = e^{-\lambda t}$ so that $\lambda = \dfrac{1}{t} \ln\!\left(\dfrac{R_0}{R}\right)$. In this case $\dfrac{R_0}{R} = 5$ when

$t = 2$ h, so $\lambda = \dfrac{1}{2\,\text{h}} \ln 5 = 0.805$ h^{-1}.

(b) $T_{1/2} = \dfrac{\ln 2}{\lambda} = \dfrac{\ln 2}{0.805 \text{ h}^{-1}} = 0.861$ h

13-23   (a)   From $R = R_0 e^{-\lambda t}$, $\lambda = \dfrac{1}{t} \ln\!\left(\dfrac{R_0}{R}\right)$, $\lambda = \dfrac{1}{4\,\text{h}} \ln\!\left(\dfrac{10}{8}\right) = 5.58 \times 10^{-2}$ h^{-1} $= 1.55 \times 10^{-5}$ s^{-1}, and

$T_{1/2} = \dfrac{\ln 2}{\lambda} = 12.4$ h.

(b) $R_0 = 10$ mCi $= 10 \times 10^{-3} \times 3.7 \times 10^{10}$ decays/s $= 3.7 \times 10^8$ decays/s and $R = \lambda N$ so

$N_0 = \dfrac{R_0}{\lambda} = \dfrac{3.7 \times 10^8 \text{ decays s}^{-1}}{1.55 \times 10^{-5} \text{ s}^{-1}} = 2.39 \times 10^{13}$ atoms

(c) $R = R_0 e^{-\lambda t} = (10 \text{ mCi})e^{-(5.58 \times 10^{-2})(30)} = 1.87$ mCi

13-25   Combining Equations 13.8 and 13.11 we have $N = \dfrac{|dN/dt|}{\lambda} = \dfrac{|dN/dt|}{0.693/T_{1/2}}$ and since

1 mCi $= 3.7 \times 10^7$ decays/s.

$$N = \frac{(5 \text{ mCi})(3.7 \times 10^7 \text{ dps/mCi})}{0.693/[(28.8 \text{ yr})(3.16 \times 10^7 \text{ s/yr})]} = 2.43 \times 10^{17} \text{ atoms}$$

Therefore, the mass of strontium in the sample is

$$m = \frac{N}{N_A} M = \frac{2.43 \times 10^{17} \text{ atoms}}{6.022 \times 10^{23} \text{ atoms/mole}}(90 \text{ g/mole}) = 36.3 \times 10^{-6} \text{ g}.$$

13-27   Let R_0 equal the total activity withdrawn from the stock solution.

$$R_0 = (2.5 \text{ mCi/ml})(10 \text{ ml}) = 25 \text{ mCi}.$$

Let R_0' equal the initial specific activity of the working solution.

$$R_0' = \frac{25 \text{ mCi}}{250 \text{ ml}} = 0.1 \text{ mCi/ml}$$

After 48 hours the specific activity of the working solution will be

$$R' = R_0' e^{-\lambda t} = (0.1 \text{ mCi/ml}) e^{-(0.693/15 \text{ h})(48 \text{ h})} = 0.011 \text{ mCi/ml}$$

and the activity in the sample will be, $R = (0.011 \text{ mCi/ml})(5 \text{ ml}) = 0.055 \text{ mCi}$.

13-29   The number of nuclei that decay during the interval will be

$$N_1 - N_2 = N_0 \left(e^{-\lambda t_1} - e^{-\lambda t_2} \right).$$

First we find λ;

$$\lambda = \frac{\ln 2}{T_{1/2}} = \frac{0.693}{64.8 \text{ h}} = 0.0107 \text{ h}^{-1} = 2.97 \times 10^{-6} \text{ s}^{-1} \text{ and}$$

$$N_0 = \frac{R_0}{\lambda} = \frac{(40 \ \mu\text{Ci})(3.7 \times 10^4 \text{ dps}/\mu\text{Ci})}{2.97 \times 10^{-6} \text{ s}^{-1}} = 4.98 \times 10^{11} \text{ nuclei}$$

Using these values we find

$$N_1 - N_2 = \left(4.98 \times 10^{11} \right) \left[e^{-(0.0107 \text{ h}^{-1})(10 \text{ h})} - e^{-(0.0107 \text{ h}^{-1})(12 \text{ h})} \right].$$

Hence, the number of nuclei decaying during the interval is

$$N_1 - N_2 = 9.46 \times 10^9 \text{ nuclei}.$$

13-31   (a)

(b) $\lambda = -\text{slope} = -\dfrac{\ln 200 - \ln 480}{(12 - 4) \text{ hr}} = 0.25 \text{ hr}^{-1} = 4.17 \times 10^{-3} \text{ min}^{-1} \text{ and } T_{1/2} = \dfrac{\ln 2}{\lambda} = 2.77 \text{ hr}.$

(c)   By extrapolation of graph to $t = 0$, we find $(\text{cpm})_0 = 4 \times 10^3 \text{ cpm}$

(d) $N = \dfrac{R}{\lambda}; \ N_0 = \dfrac{R_0}{\lambda} = \dfrac{(\text{cpm})_0 / \text{EFF}}{\lambda}$

$N_0 = \dfrac{4 \times 10^4 \text{ dis/min}}{4.17 \times 10^{-3} \text{ min}^{-1}} = 9.59 \times 10^6 \text{ atoms}$

13-33   (a)   Referring to Example 13.11 or using the note in Problem 35 $R = R_0 e^{-\lambda t}$,

$$R_0 = N_0 \lambda = 1.3 \times 10^{-12} N_0 \left(^{12}C\right) \lambda$$

$$R_0 = \left(\frac{1.3 \times 10^{-12} \times 25 \text{ g} \times 6.02 \times 10^{23} \text{ atoms/mole}}{12 \text{ g/mole}} \right) \lambda$$

where $\lambda = \dfrac{0.693}{5\,730 \times 3.15 \times 10^7} = 3.84 \times 10^{-12}$ decay/s. So $R_0 = 376$ decay/min, and

$$R = \left(3.76 \times 10^2\right) \exp\left[\left(-3.84 \times 10^{-12} \text{ s}^{-1}\right) \times \left(2.3 \times 10^4 \text{ y}\right) \times \left(3.15 \times 10^7 \text{ s/y}\right)\right]$$
$$R = 18.3 \text{ counts/min}$$

(b)   The observed count rate is slightly less than the average background and would be difficult to measure accurately within reasonable counting times.

13-35   First find the activity per gram at time $t = 0$, $R_0 = N_0\left(^{14}C\right)$, where

$N_0\left(^{14}C\right) = 1.3 \times 10^{-12} N_0\left(^{12}C\right)$; and $N_0\left(^{12}C\right) = \left(\dfrac{m}{M}\right)N_a$. Therefore $\dfrac{R_0}{m} = \left(\dfrac{\lambda N_a}{M}\right)\left(1.3 \times 10^{-12}\right)$ and

the activity after decay at time t will be $\dfrac{R}{m} = \left(\dfrac{R_0}{m}\right)e^{-\lambda t} = \left(\dfrac{\lambda N_a}{M}\right)\left(1.3 \times 10^{-12}\right)e^{-\lambda t}$ where

$\lambda = \dfrac{\ln 2}{T_{1/2}} = 2.3 \times 10^{-10}$ min^{-1} when $t = 2\,000$ years.

$$\frac{R}{m} = \left(\frac{3.2 \times 10^{-10} \text{ min}^{-1}}{12 \text{ g/mole}} \right)\left(1.3 \times 10^{-12}\right)\left(6.03 \times 10^{23} \text{ mole}^{-1}\right) \times e^{-\left(3.2 \times 10^{-10} \text{ min}^{-1}\right)\left(2\,000 \text{ y}\right)\left(5.26 \times 10^5 \text{ min/y}\right)}$$

$$\frac{R}{m} = 11.8 \text{ decays min}^{-1}\text{g}^{-1}$$

13-37   (a)   Let N_1 = number of parent nuclei, and N_2 = number of daughter nuclei. The daughter nuclei increase at the rate at which the parent nuclei decrease, or

$$\frac{dN_2}{dt} = \frac{-dN_1}{dt} = \lambda N_1 = \lambda N_{01} e^{-\lambda_1 t}$$
$$dN_2 = \lambda N_{01} e^{-\lambda_1 t} dt$$
$$N_2 = \lambda N_{01} \int e^{-\lambda t} dt = -N_{01} e^{-\lambda t} + \text{Const.}$$

If we require $N_2 = N_{02}$ when $t = 0$ then Const $= N_{02} + N_{01}$. Therefore $N_2 = N_{02} + N_{01} - N_{01}e^{-\lambda t}$. And when $N_{02} = 0$; $N_2 = N_{01}\left(1 - e^{-\lambda t}\right)$.

(b)   Obtain the number of parent nuclei from $N_1 = N_{01}e^{-\lambda t}$ and the daughter nuclei from $N_2 = N_{01}\left(1 - e^{-\lambda t}\right)$ with $N_{01} = 10^6$, $\lambda = \dfrac{\ln 2}{T_{1/2}} = \dfrac{0.693}{10 \text{ h}} = 0.069\,3$ h^{-1}. Thus the quantities

$N_1 = 10^6 e^{-(0.069\,3 \text{ h}^{-1})t}$ and $N_2 = 10^6\left[1 - e^{-(0.069\,3 \text{ h}^{-1})t}\right]$ are plotted below.

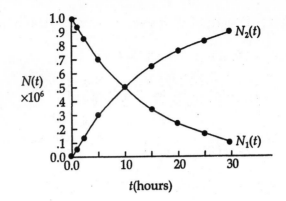

13-39   A number of atoms, $dN = \lambda N dt$, have life times of t. Therefore, the average or mean life time will be $\sum (dN) \dfrac{t}{\sum dN}$ or $\int dN \dfrac{t}{N_0}$ so $\tau = \dfrac{1}{N_0} \int\limits_0^\infty \lambda N t \, dt = \dfrac{1}{N_0} \int\limits_0^\infty \lambda N_0 e^{-\lambda t} t \, dt = \dfrac{1}{\lambda}$.

13-41 $Q = \left(M_{238_{U}} - M_{234_{Th}} - M_{4_{He}} \right)(931.5 \text{ MeV/u})$

$= (238.048\,608 \text{ u} - 234.043\,583 \text{ u} - 4.002\,603 \text{ u})(931.5 \text{ MeV/u}) = 2.26 \text{ MeV}$

13-43   (a)   We will assume the parent nucleus (mass M_p) is initially at rest, and we will denote the masses of the daughter nucleus and alpha particle by M_d and M_a, respectively. The equations of conservation of momentum and energy for the alpha decay process are

$$M_d v_d = M_\alpha v_\alpha \tag{1}$$

$$M_p c^2 = M_d c^2 + M_\alpha c^2 + \left(\frac{1}{2}\right) M_d v_d^2 + \left(\frac{1}{2}\right) M_\alpha v_\alpha^2 \tag{2}$$

The disintegration energy Q is given by

$$Q = \left(M_p - M_d - M_\alpha \right) c^2 = \left(\frac{1}{2}\right) M_d v_d^2 + \left(\frac{1}{2}\right) M_\alpha v_\alpha^2 \tag{3}$$

Eliminating v_d from Equations (1) and (3) gives

$$Q = \left(\frac{1}{2}\right) M_d \left[\left(\frac{M_\alpha}{M_d} \right) v_\alpha \right]^2 + \left(\frac{1}{2}\right) M_\alpha v_\alpha^2$$

$$Q = \left(\frac{1}{2}\right) \left(\frac{M_\alpha^2}{M_d} \right) v_\alpha^2 + \left(\frac{1}{2}\right) M_\alpha v_\alpha^2$$

$$Q = \left(\frac{1}{2}\right) M_\alpha v_\alpha^2 \left(1 + \frac{M_\alpha}{M_d} \right) = K_\alpha \left(1 + \frac{M_\alpha}{M_d} \right)$$

(b) $K_\alpha = \dfrac{Q}{1 + M_\alpha / M} = \dfrac{4.87 \text{ MeV}}{1 + 4/226} = 4.79 \text{ MeV}$

(c) $K_d = (4.87 - 4.97) \text{ MeV} = 0.08 \text{ MeV}$

(d)   For the beta decay of ^{210}Bi we have $Q = K_{e^-}\left(1 + \dfrac{M_{e^-}}{M_Y}\right)$. Solving for K_{e^-} and

substituting $M_{e^-} = 5.486 \times 10^{-4}$ u and $M_Y = 209.982$ u (Po), we find

$$K_{e^-} = \frac{Q}{1 + 5.486 \times 10^{-4}\ \text{u}/209.982\ \text{u}} = \frac{Q}{1 + 2.61 \times 10^{-6}}.$$

Setting $2.61 \times 10^{-6} = \varepsilon$, we get $K_{e^-} = Q(1+\varepsilon)^{-1} \cong Q(1-\varepsilon) = Q(1 - 2.61 \times 10^{-6})$ for $\varepsilon \ll 1$. This means the daughter Po carries off only about three millionths of the kinetic energy available in the decay. This treatment is only approximately correct since actual beta decay involves another particle (antineutrino) and relativistic effects.

13-45 $Q = (m_{\text{initial}} - m_{\text{final}})(931.5\ \text{MeV/u})$

(a) $Q = m\binom{40}{20}\text{Ca} - m(e^+) - m\binom{40}{19}\text{K}) = (39.962\,59\ \text{u} - 0.000\,548\,6\ \text{u} - 39.964\,00\ \text{u})(931.5\ \text{MeV/u})$
       $= -1.82\ \text{MeV}$
       $Q < 0$ so the reaction cannot occur.

(b)   Using the handbook of Chemistry and Physics
       $Q = m\binom{98}{44}\text{Ru}) - m\binom{4}{2}\text{He}) - m\binom{94}{42}\text{Mo}) = (97.905\,5\ \text{u} - 4.002\,6\ \text{u} - 93.904\,7\ \text{u})(931.5\ \text{MeV/u})$
       $= -1.68\ \text{MeV}$
       $Q < 0$ so the reaction cannot occur.

(c)   Using the handbook of *Chemistry and Physics*
       $Q = m\binom{144}{60}\text{Nd}) - m\binom{4}{2}\text{He}) - m\binom{140}{58}\text{Ce}) = (143.909\,9\ \text{u} - 4.002\,6\ \text{u} - 139.905\,4\ \text{u})$
       $\times (931.5\ \text{MeV/u}) = 1.86\ \text{MeV}$
       $Q > 0$ so the reaction can occur.

13-47   We assume an electron in the nucleus with an uncertainty in its position equal to the nuclear diameter. Choose a typical diameter of 10 fm and from the uncertainty principle we have

$$\Delta p \approx \frac{h}{\Delta x} = 6.6 \times 10^{-34}\ \text{J s}/10^{-14}\ \text{m} = 6.6 \times 10^{-20}\ \text{N s}.$$

Using the relativistic energy-momentum expression

$$E^2 = (pc)^2 + (m_0 c^2)^2$$

we make the approximation that $pc \approx (\Delta p)c \gg m_0 c^2$ so that

$$E \approx pc \approx (\Delta p)c = (6.6 \times 10^{-20}\ \text{N s})(3 \times 10^8\ \text{m/s}) = 19.8 \times 10^{-12}\ \text{J} \approx 124\ \text{MeV}.$$

However, the most energetic electrons emitted by radioactive nuclei have been found to have energies of less than 10% of this value, therefore electrons are not present in the nucleus.

13-49   The disintegration energy, Q, is c^2 times the mass difference between the parent nucleus and the decay products. In electron emission an electron leaves the system. That is $^A_Z X \rightarrow ^A_{Z+1} Y + e^- + \overline{v}$ where $\overline{v}$ has negligible mass and the neutral daughter nucleus has nuclear charge of $Z+1$ and Z electrons. Therefore we need to add the mass of an electron to get the mass of the daughter. The disintegration energy can now be calculated as

$$Q = \{M^A_Z X - M[^A_{Z+1} Y - m_e] - m_e + 0\}c^2 = [M^A_Z X - M^A_{Z+1} Y]c^2.$$

Similar reasoning can be applied to positron emission $^A_Z X \rightarrow ^A_{Z-1} Y + e^+ + v$ and so

$$Q = \{M^A_Z X - M[^A_{Z-1} Y - m_e] - m_e + 0\}c^2 = [M^A_Z X - M^A_{Z-1} Y - 2m_e]c^2.$$

For electron capture we have $^A_Z X + e^- \rightarrow ^A_{Z+1} Y + v$, which gives

$$Q = \{M^A_Z X + m_e - M[^A_{Z-1} Y + m_e] + 0\}c^2 = [M^A_Z X - M^A_{Z-1} Y]c^2.$$

13-51   In the decay $^3_1 H \rightarrow ^3_2 He + e + \overline{v}$ the energy released is: $E = (\Delta m)c^2 = [M_{1H} - M_{3He}]c^2$ since the mass of the antineutrino is negligible and the mass of the electron is accounted for in the atomic masses of $^3_1 H$ and $^3_2 He$. Thus,

$$E = (3.016\ 049\ u - 3.016\ 029\ u)(931.5\ MeV/u) = 0.018\ 6\ MeV = 18.6\ keV.$$

13-53 $$N_{Rb} = 1.82 \times 10^{10} \left(^{87}Rb\ atoms/g\right)$$

$$N_{Sr} = 1.07 \times 10^9 \left(^{87}Sr\ atoms/g\right)$$

$$T_{1/2}\left(^{87}Rb \xrightarrow{\beta^-} {}^{87}Sr\right) = 4.8 \times 10^{10}\ y$$

(a)   If we assume that all the ^{87}Sr came from ^{87}Rb, then $N_{Rb} = N_0 e^{-\lambda t}$

$$1.82 \times 10^{10} = \left(1.82 \times 10^{10} + 1.07 \times 10^9\right) e^{-(\ln 2/4.8 \times 10^{10})t}$$

$$-\ln(0.944\ 47) = \left(\frac{\ln 2}{4.8 \times 10^{10}}\right)t$$

$$t = 3.96 \times 10^9\ y$$

(b)   It could be no older. The rock could be younger if some ^{87}Sr were initially present.

13-55   (a)   Starting with $N = 0$ radioactive atoms at $t = 0$, the rate of increase is (production-decay)

$$\frac{dN}{dt} = R - \lambda N$$

$$dN = (R - \lambda N)dt$$

Variables are separable

$$\int_{N=0}^{N} \frac{dN}{R-\lambda N} = \int_{t=0}^{t} dt - \left(\frac{1}{\lambda}\right)\ln\left(\frac{R-\lambda N}{R}\right) = t$$

$$\ln\left(\frac{R-\lambda N}{R}\right) = -\lambda t$$

$$\left(\frac{R-\lambda N}{R}\right) = e^{-\lambda t}$$

$$1 - \left(\frac{\lambda}{R}\right)N = e^{-\lambda t}$$

$$N = \left(\frac{R}{\lambda}\right)\left(1 - e^{-\lambda t}\right)$$

(b) $\dfrac{dN}{dt} = R - \lambda N_{max}$

$N_{max} = \dfrac{R}{\lambda}$

13-57   We have all this information: $N_x(0) = 2.50 N_y(0)$

$$N_x(3d) = 4.20 N_y(3d)$$

$$N_x(0)e^{-\lambda_x 3d} = 4.20 N_y(0)e^{-\lambda_y 3d} = 4.20 \frac{N_x(0)}{2.50}e^{-\lambda_y 3d}$$

$$e^{3d\lambda_x} = \frac{2.5}{4.2}e^{3d\lambda_y}$$

$$3d\lambda_x = \ln\left[\frac{2.5}{4.2}\right] + 3d\lambda_y$$

$$3d\frac{0.693}{T_{1/2x}} = \ln\left(\frac{2.5}{4.2}\right) + 3d\frac{0.693}{1.60 \text{ d}} = 0.781$$

$$T_{1/2x} = 2.66 \text{ d}$$

13-59 $N = N_0 e^{-\lambda t}$

$$\left|\frac{dN}{dt}\right| = R = \left|-\lambda N_0 e^{-\lambda t}\right| = R_0 e^{-\lambda t}$$

$$e^{-\lambda t} = \frac{R}{R_0}$$

$$e^{\lambda t} = \frac{R_0}{R}$$

$$\lambda t = \ln\left(\frac{R_0}{R}\right) = \frac{\ln 2}{T_{1/2}}t$$

$$t = T_{1/2}\frac{\ln(R_0/R)}{\ln 2}$$

If $R = 0.13$ Bq, $t = 5\,730 \text{ yr}\dfrac{\ln(0.25/0.13)}{0.693} = 5\,406 \text{ yr}$.

If $R = 0.11$ Bq, $t = 5\,730 \text{ yr}\dfrac{\ln(0.25/0.11)}{0.693} = 6\,787 \text{ yr}$.

The range is most clearly written as between 5 400 yr and 6 800 yr, without understatement.

13-61   (a)   Let N be the number of ^{238}U nuclei and N' be ^{206}Pb nuclei. Then $N = N_0 e^{-\lambda t}$ and $N_0 = N + N'$ so $N = (N + N')e^{-\lambda t}$ or $e^{\lambda t} = 1 + \dfrac{N'}{N}$. Taking logarithms, $\lambda t = \ln\left(1 + \dfrac{N'}{N}\right)$ where $\lambda = \dfrac{\ln 2}{T_{1/2}}$. Thus, $t = \left(\dfrac{T_{1/2}}{\ln 2}\right)\ln\left(1 + \dfrac{N'}{N}\right)$. If $\dfrac{N}{N'} = 1.164$ for the ^{238}U $\rightarrow$ ^{206}Pb chain with $T_{1/2} = 4.47 \times 10^9$ yr, the age is:

$$t = \left(\frac{4.47 \times 10^9 \text{ yr}}{\ln 2}\right)\ln\left(1 + \frac{1}{1.164}\right) = 4.00 \times 10^9 \text{ yr}.$$

(b)   From above, $e^{\lambda t} = 1 + \dfrac{N'}{N}$. Solving for $\dfrac{N}{N'}$ gives $\dfrac{N}{N'} = \dfrac{e^{-\lambda t}}{1 - e^{-\lambda t}}$. With $t = 4.00 \times 10^9$ yr and $T_{1/2} = 7.04 \times 10^8$ yr for the ^{235}U $\rightarrow$ ^{207}Pb chain,

$$\lambda t = \left(\frac{\ln 2}{T_{1/2}}\right)t = \frac{(\ln 2)(4.00 \times 10^9 \text{ yr})}{7.04 \times 10^8 \text{ yr}} = 3.938 \text{ and } \frac{N}{N'} = 0.019\,9.$$

With $t = 4.00 \times 10^9$ yr and $T_{1/2} = 1.41 \times 10^{10}$ yr for the ^{232}Th $\rightarrow$ ^{208}Pb chain,

$$\lambda t = \frac{(\ln 2)(4.00 \times 10^9 \text{ yr})}{1.41 \times 10^{10} \text{ yr}} = 0.196\,6 \text{ and } \frac{N}{N'} = 4.60.$$

14

Nuclear Physics Applications

14-1 $^{18}O = 17.999\ 160$ $^{18}F = 18.000\ 938$

      $m_n = 1.008\ 664\ 9$ $^{1}H = 1.007\ 825$ all in u.

    (a) $Q = [M_O + M_H + M_F - m_n]c^2 = [-0.002\ 617\ 9\ u][931.494\ 3\ MeV/u] = -2.438\ 6\ MeV$
          compared to $-2.453 \pm 0.000\ 2\ MeV$.

    (b) $K_{th} = -Q\left[1 + \dfrac{M_a}{M_\chi}\right] = (2.438\ 6\ MeV)\left(1 + \dfrac{1.007\ 825}{17.999\ 160}\right) = 2.575\ 1\ MeV$

14-3 $Q = \left(M_\alpha + M_{(^9Be)} - M_{(^{12}C)} - M_n\right)(931.5\ MeV/u)$

      $= (4.002\ 603\ u + 9.012\ 182\ u - 12.000\ 000\ u - 1.008\ 665\ u)(931.5\ MeV/u)$

    $Q = 5.70\ MeV$

14-5 $Q = (m_a + m_X - m_Y - m_b)[931.5\ MeV/u]$

    $Q = \left[m_{(^1H)} + m_{(^7Li)} - m_{(^4He)} - m_\alpha\right]u[931.5\ MeV/u]$

    $Q = [1.007\ 825\ u + 7.016\ 004\ u - 4.002\ 603\ u - 4.002\ 603\ u][931.5\ MeV/u]$

    $Q = 17.35\ MeV$

14-7   (a) $Q = \left[m(^{14}N) + m(^4He) - m(^{17}O) - m(^1H)\right](931.5\ MeV/u)$
          Using Table 13.6 for the masses.

        $Q = (14.003\ 074\ u + 4.002\ 603\ u - 16.999\ 132\ u - 1.007\ 825\ u)(931.5\ MeV/u)$

        $Q = -1.19\ MeV$

        $K_{th} = -\dfrac{Q\left[m(^4He) + m(^{14}N)\right]}{m(^{14}N)} = -(-1.19\ MeV)\left(1 + \dfrac{4.002\ 603}{14.003\ 074}\right) = 1.53\ MeV$

    (b) $Q = \left[m(^7Li) + m(^1H) - 2m(^1He)\right](931.5\ MeV/u)$

        $Q = [(7.016\ 004\ u + 1.007\ 825\ u) - (2)(4.002\ 603\ u)](931.5\ MeV/u)$

        $Q = 17.35\ MeV$

14-9   (a)   <u>CM SYSTEM</u>

$$p = M_a v = M_\chi V$$

$$K_{CM} = \frac{p^2}{2M_a} + \frac{p^2}{2M_\chi} = \frac{p^2}{2}\left[\frac{M_\chi + M_a}{M_a M_\chi}\right]$$

<u>LAB SYSTEM</u>

$$P_{lab} = M_a(v+V)(\text{Eq. 1})$$

$$= p\left[\frac{M_\chi + M_a}{M_\chi}\right] \text{ for substituting } v = \frac{p}{M_a} \text{ and } V = \frac{p}{M_\chi} \text{ in Eq. 1.}$$

$$K_{lab} = \frac{p_{lab}^2}{2M_a} = \frac{p^2\left[(M_\chi + M_a)/M_\chi\right]^2}{2M_a}$$

Comparing to K_{CM}, $K_{lab} = K_{CM}\left[\dfrac{M_\chi + M_a}{M_\chi}\right]$ or $K_{th} = -Q\left(1 + \dfrac{M_a}{M_\chi}\right)$

(b)   First calculate the Q-value

$$Q = \left[m(^{14}\text{N}) + m(^4\text{He}) - m(^{17}\text{O}) - m(^1\text{H})\right](931.5 \text{ MeV/u})$$

$$Q = [14.003\,074 \text{ u} + 4.002\,603 \text{ u} - 16.999\,132 \text{ u} - 1.007\,825 \text{ u}](931.5 \text{ MeV/u})$$

$$Q = -1.19 \text{ MeV}$$

Then

$$K_{th} = -Q\left[1 + \frac{m(^4\text{He})}{m(^{14}\text{N})}\right]$$

$$K_{th} = -(-1.19 \text{ MeV})\left[1 + \frac{4.002\,603}{14.003\,074}\right] = 1.53 \text{ MeV}$$

14-11 $R = R_0 e^{-n\sigma x}$, $x = 2$ m, $R = 0.8R_0$, $n = \dfrac{\rho}{m_{atom}} = \dfrac{70 \text{ kg/m}^3}{1.67 \times 10^{-27} \text{ kg}} = 4.19 \times 10^{28} \text{ m}^{-3}$, $0.8R_0 = R_0 e^{-n\sigma x}$,

$0.8 = e^{-n\sigma x}$, $n\sigma x = -\ln 0.8$, $\sigma = \dfrac{-1}{nx}\ln(0.8) = \dfrac{0.223}{4.19 \times 10^{28} \text{ m}^{-3} \times 2 \text{ m}} = 2.66 \times 10^{-30} \text{ m}^2 = 0.026\,6 \text{ b}$

14-13 Equation 14.4 gives $R = (R_0 n x)\sigma$. Using values of E and σ, we have

(a) $\dfrac{R_{10}}{R_1} = \dfrac{\sigma_{10}}{\sigma_1} = 0.037\,3$,

(b) $\dfrac{R_1}{R_{0.1}} = 0.066\,3$, and

(c) $\dfrac{R_{0.1}}{R_{0.01}} \approx 1$

(d) Therefore we can use cadmium as an energy selector in the range 0.1 eV to 10 eV to detect order of magnitude changes in energy.

14-15 (a) $\dfrac{N}{N_0} = e^{-n\sigma x}$, $x =$ thickness in m, $\sigma =$ cross section in m^2 and

$$n = \#\ \text{gold nuclei}/m^3$$
$$n = (6.02 \times 10^{23}\ \text{atoms/mole})(1\ \text{mole}/197\ \text{g})(19.3\ \text{g}/cm^3)$$
$$n = 5.9 \times 10^{22}\ \text{atoms}/cm^3 = 5.9 \times 10^{28}\ \text{atoms}/m^3$$

Taking $x = 5.1 \times 10^{-5}$ m, we get

$$\frac{N}{N_0} = \exp(-5.9 \times 10^{28}\ \text{atoms}/m^3 \times 500 \times 10^{-28}\ m^2 \times 5.1 \times 10^{-5}\ m) = 0.86$$

(b) $N = 0.86 N_0 \qquad N_0 = \dfrac{0.1\ \mu A}{1.6 \times 10^{-19}\ C}$

$N_0 = 6.3 \times 10^{11}$ protons/s $\qquad$ and $\qquad N = 6.1 \times 10^{11}$ protons/s

(c) The number of protons abs. or scat. per sec $0.14 N_0 = 8.7 \times 10^{10}$ protons/s

14-17 Since $N = N_0 e^{-n\sigma x}$, $\dfrac{dN}{dx} = -N n_c \sigma$, where $N =$ neutron density, $n_c =$ cadmium nuclei density, and σ is the absorption cross-section. Thus, $\left(\dfrac{dN}{dt}\right)_a = -N n_c \sigma v_{th}$ where v_{th} is the neutron thermal velocity given by $v_{th} = \left(\dfrac{1.5 k_B T}{m_n}\right)^{1/2}$. The neutron decay rate, $\left(\dfrac{dN}{dt}\right)_D$, comes from differentiating $N = N_0 e^{-\lambda t}$: $\left(\dfrac{dN}{dt}\right)_D = -N\lambda$ where $\lambda = \dfrac{0.693}{T_{1/2}} = \dfrac{0.693}{636\ \text{s}} = 1.09 \times 10^{-3}$ s^{-1}. Finally

$$\frac{(dN/dt)_a}{(dN/dt)_D} = \frac{-N n_c \sigma v_{th}}{-N\lambda} = \frac{n_c \sigma v_{th}}{\lambda}$$

$$\text{As } n_c = (8.65 \text{ g/cm}^3)(6.02 \times 10^{23} \text{ nuclei/112 g})$$

$$\sigma = (2\,450 \text{ b})(10^{-24} \text{ cm}^2/\text{b})$$

$$v_{th} = \left[\frac{(1.5)(1.38 \times 10^{-23} \text{ J/K})(300 \text{ K})}{1.67 \times 10^{-27} \text{ kg}} \right]^{1/2}$$

$$\lambda = 1.09 \times 10^{-3} \text{ s}^{-1}$$

$$\frac{(dN/dt)_a}{(dN/dt)_D} = 2.25 \times 10^{12}$$

14-19 $E_T \equiv E(\text{thermal}) = \frac{3}{2} k_B T = 0.038\,9 \text{ eV}$. $E_T = \left(\frac{1}{2}\right)^n E$ where $n \equiv$ number of collisions, and E is the

initial kinetic energy. $0.0389 = \left(\frac{1}{2}\right)^n (10^6)$. Therefore $n = 24.6$ or 25 collisions.

14-21 $\Delta E = c^2(m_U - m_{Ba} - m_{Kr} - m_n)$
$\Delta E = (931.5 \text{ MeV/u})[235.043\,9 \text{ u} - 140.913\,9 \text{ u} - 91.897\,3 \text{ u} - 2(1.008\,7 \text{ u})]$
$\Delta E = (931.5 \text{ MeV/u})[0.215\,3 \text{ u}] = 200.6 \text{ MeV}$

14-23 (a)   For a sphere: $V = \frac{4}{3}\pi r^3$ and $r = \left(\frac{3V}{4\pi}\right)^{1/3}$, so $\frac{A}{V} = \frac{4\pi r^2}{(4/3)\pi r^3} = 4.84 V^{-1/3}$.

(b)   For a cube: $V = l^3$ and $l = V^{1/3}$, so $\frac{A}{V} = \frac{6l^2}{l^3} = 6V^{-1/3}$.

(c)   For a parallelepiped: $V = 2a^3$ and $a = (2V)^{1/3}$, so $\frac{A}{V} = \frac{2a^2 + 8a^2}{2a^3} = 6.30 V^{-1/3}$.

(d)   Therefore for a given volume, the sphere has the least leakage.

(e)   The parallelepiped has the greatest leakage.

14.25 (a) $\text{eff} = \frac{P_{delivered}}{P_{out}} = 0.3$, $P_{out} = \frac{1\,000 \text{ MW}}{0.3} = 3\,333 \text{ MW}$

(b) $P_{heat} = P_{out} - P_{delivered} = 3\,333 - 1\,000 = 2\,333 \text{ MW}$

(c)   The energy released per fission event is $Q = 200 \text{ MeV}$. Therefore

$$\text{Rate} = \frac{P_{out}}{Q} = \frac{3.333\,3 \times 10^9 \text{ W}/200 \text{ MeV}}{1.6 \times 10^{-13} \text{ J/MeV}}$$

$$\text{Rate} = 1.04 \times 10^{20} \text{ events/s}$$

(d) $M = (\text{Rate}) \left[\dfrac{235 \times 10^{-3} \text{ kg/mole}}{6.0 \times 10^{23} \text{ atoms/mole}} \right] (\text{time})$

$M = (1.04 \times 10^{20} \text{ events/s})(3.92 \times 10^{-25} \text{ kg/atom})(365 \text{ days})(24 \text{ h/day}) \times (3\,600 \text{ s/h})$

$\quad = 1.34 \times 10^3 \text{ kg}$

(e) $\dfrac{dM}{dt} = \left(\dfrac{1}{c^2} \right) \left(\dfrac{dE}{dt} \right) = \dfrac{3.333 \times 10^9 \text{ W}}{(3 \times 10^8 \text{ m/s})^2} = 3.7 \times 10^{-8} \text{ kg/s}$. To compare with (d) we need the

mass for a year.

$\quad \dfrac{dM}{dt}(\text{year}) = (3.7 \times 10^{-8} \text{ kg/s})(365 \text{ days})(24 \text{ h/day}) \times (3\,600 \text{ s/h}) = 1.17 \text{ kg}.$

This is 8% of the total mass found in (d).

14-27   (a) $r = r_D + r_T = (1.2 \times 10^{-15} \text{ m})(2^{1/3} + 3^{1/3}) = 2.70 \times 10^{-15} \text{ m}$

(b) $U = \dfrac{ke^2}{r} = \dfrac{(9 \times 10^9 \text{ N m}^2/\text{C}^2)(1.6 \times 10^{-19} \text{ C})^2}{2 \times 10^{-15} \text{ m}} = 1.15 \times 10^{-13} \text{ J} = 720 \text{ keV}$

(c)   Conserving momentum: $v_F = \dfrac{v_0 m_D}{m_D + m_T}$ (1)

(d) $\dfrac{1}{2} m_D v_0^2 = \dfrac{1}{2}(m_D + m_T)v_F^2 + U$ (2)

Eliminating v_F from (2) using (1), gives

$\quad \left(\dfrac{m_D}{2} \right) v_0^2 - \dfrac{1}{2}(m_D + m_T) \dfrac{v_0^2 m_D^2}{(m_D + m_T)^2} = U$ or

$\quad \dfrac{1}{2}(m_D + m_T)m_D v_0^2 - \dfrac{1}{2} m_D^2 v_0^2 = (m_D + m_T)U$ or

$\quad\quad \dfrac{1}{2} m_D^2 v_0^2 = \left(\dfrac{m_D + m_T}{m_T} \right) U = \dfrac{5}{3} U = \dfrac{5}{3}(720 \text{ keV})$

$\quad\quad \dfrac{1}{2} m_D^2 v_0^2 = 1.2 \text{ MeV}.$

(e)   Possibly by tunneling.

14-29   (a) $Q = K_\alpha + K_n = 17.6 \text{ MeV} = (1.2)m_\alpha v_\alpha^2 + \dfrac{1}{2} m_n v_n^2$. Momentum conservation yields

$m_n v_n = m_\alpha v_\alpha$. Substituting $v_\alpha = \dfrac{m_n}{m_\alpha} v_n$ into the energy equation gives $K_n = \dfrac{m_\alpha Q}{m_\alpha + m_n}$,

$K_\alpha = \dfrac{m_n Q}{m_\alpha + m_n}$ Finally, $K_n = \dfrac{(4.003)(17.6 \text{ MeV})}{4.003 + 1.009} = 14.1 \text{ MeV}$, $K_\alpha = 3.45 \text{ MeV}.$

(b)   Yes, since the neutron is uncharged, it is not confined by the **B** field and only K_α can be used to achieve critical ignition.

14-31   (a)   The pellet contains

$$\left(\frac{4\pi R^3}{3}\right)(0.2 \text{ g/cm}^3) = \left(\frac{4\pi(0.5\times10^{-2} \text{ cm})^3}{3}\right)(0.2 \text{ g/cm}^3) = 1.05\times10^{-7} \text{ g}$$

of $_1^2\text{H} + _1^3\text{H}$ "molecules." The number of molecules, N, is

$$\left(\frac{1.05\times10^{-7} \text{ g}}{5.0 \text{ g/mole}}\right)(6.02\times10^{23} \text{ molecules/mole}) = 1.26\times10^{16}.$$

Since each molecule consists of 4 particles $\left(_1^2\text{H}, _1^3\text{H}, 2e^-\right)$, $E = (4N)\frac{3}{2}k_BT$ or

$$T = \frac{E}{6Nk_B} = \frac{0.01(200\times10^3 \text{ J})}{6(1.26\times10^{16})(1.38\times10^{-23} \text{ J/K})} = 1.9\times10^9 \text{ K}.$$

(b)   The energy released $= (17.59 \text{ MeV})(1.26\times10^{16})(1.6\times10^{-13} \text{ J/MeV}) = 355 \text{ kJ}.$

14-33   (a)   Roughly $\frac{7}{2}(15\times10^6 \text{ K})$ or 52×10^6 K since 6 times the coulombic barrier must be surmounted.

(b) $Q = \Delta mc^2 = (12.000\,000 \text{ u} + 1.007\,825 \text{ u} - 13.005\,738 \text{ u})(931.5 \text{ MeV/u})$
$Q = 1.943$ MeV
The other energies are calculated in a similar manner and the total energy released is

$$(1.943 + 1.709 + 7.551 + 7.297 + 2.242 + 4.966) \text{ MeV} = 25.75 \text{ MeV}.$$

The net effect is $_6^{12}\text{C} + 4p \rightarrow _6^{12}\text{C} + _2^4\text{He}.$

(c)   Most of the energy is lost since v's have such low cross-section (no charge, little mass, etc.)

14-35   Total energy $=$ number of ^{6}Li nuclei (22 MeV)

$$= (0.075)(2\times10^{-13} \text{ g})\left(\frac{6.02\times10^{23} \text{ nuclei}}{6.01 \text{ g}}\right)(22 \text{ MeV})(1.60\times10^{13} \text{ J/MeV}) = 5.3\times10^{23} \text{ J}$$

About twice as great as total world's fuel supply.

14-37   (a) $N =$ number of $_1^3\text{H}$,

$$_1^2\text{H pairs in 3 mg} = \frac{(3\times10^{-3} \text{ g})(6.02\times10^{23} \text{ pairs/mole})}{5.0 \text{ g/mole}}$$
$$= 3.61\times20^{20} \text{ pairs}.$$

$$\text{Power Output} = (10)(0.3)(3.61\times10^{20})(17.6 \text{ MeV/fusion})(1.60\times10^{-13} \text{ J/MeV})\big/\text{s}$$
$$= 3.1\times10^9 \text{ W}$$
$$\text{Power Input} = (10)(5\times10^{14} \text{ J/s})(10^{-8} \text{ s})\big/\text{s} = 5\times10^7 \text{ W}$$
$$\text{Net Power} = (3.1\times10^9 - 5\times10^7)\text{W} \approx 3.0\times10^9 \text{ W} = 3\,000 \text{ MW}$$

(b)   1 day's fusion energy $= (3\,000 \text{ MW})(3\,600 \text{ s/h})(24 \text{ h/day}) = 2.6 \times 10^{14}$ J. This is

equivalent to $\dfrac{2.6 \times 10^{14} \text{ J}}{50 \times 10^6 \text{ J/liter}} = 5.2 \times 10^6$ liters of oil or 5 million liters of oil!

14-39   (a) $E = \dfrac{ke^2 Z_1 Z_2}{r} = \dfrac{(9 \times 10^9 \text{ N m}^2/\text{C}^2)(1.6 \times 10^{-19} \text{ C})^2 Z_1 Z_2}{10^{-14} \text{ m}} = 2.3 \times 10^{-19} Z_1 Z_2$ J

(b)   D-D and D-T: $Z_1 = Z_2 = 1$ and $E = 2.3 \times 10^{-19}$ J $= 0.14$ MeV

14-41   (a) $E = (931.5 \text{ MeV/u})\Delta m = (931.5 \text{ MeV/u})[(2 \times 2.014\,102) - 4.002\,603]$ u. $E = 23.85$ MeV for every two ^{2}H's.

$(3.17 \times 10^8 \text{ mi}^3)[(5\,280 \text{ ft/mi})(12 \text{ in/ft})(0.025\,4 \text{ m/in})]^3 [10^6 \text{ g(H}_2\text{O)/m}^3] \left[\dfrac{2 \text{ g(H)}}{18 \text{ g(H}_2\text{O)}} \right]$

$\times [6.02 \times 10^{23} \text{ protons/g(H)}](0.015\,6 \ ^2\text{H/proton})(23.85 \text{ MeV}/^2\text{H})(1.6 \times 10^{-13} \text{ J/MeV})$

$= 2.63 \times 10^{33}$ J

(b) $\left(\dfrac{2.63 \times 10^{33} \text{ J}}{7 \times 10^{14} \text{ J/s}} \right)\left(\dfrac{\text{year}}{3.16 \times 10^7 \text{ s}} \right) = 119$ billion years

14-43   (a) $n = \dfrac{10^{14} \text{ s/cm}^3}{1 \text{ s}} = 10^{14}/\text{cm}^3$

(b) $2nk_B T = (2 \times 10^{14}/\text{cm}^3)(1.38 \times 10^{-23} \text{ J/K})(8 \times 10^7 \text{ K})(10^6 \text{ cm}^3/\text{m}^3)$

$2nk_B T = 2.2 \times 10^5 \text{ J/m}^3$

(c) $\dfrac{B^2}{2\mu_0} \approx 10(2nk_B T)$ $B = [20\mu_0(2nk_B T)]^{1/2}$

$B = [20(4\pi \times 10^{-7} \text{ N/A}^2)(2.2 \times 10^5 \text{ J/m}^3)]^{1/2} = 2.35$ T

14-45   (a)   For the first layer: $I_1 = I_0 e^{-(\mu_{Al} d)}$, for the second layer: $I_2 = I_1 e^{-(\mu_{Cu} d)}$, and for the third

layer: $\dfrac{I_0}{3} = I_2 e^{-(\mu_{Pb})d}$ so that $\dfrac{I_0}{3} = I_0 e^{-d(\mu_{Al} + \mu_{Cu} + \mu_{Pb})}$. Using Table 14.2,

$d = \dfrac{\ln 3}{\mu_{Al} + \mu_{Cu} + \mu_{Pb}} = \dfrac{\ln 3}{(5.4 + 170 + 610)(\text{cm}^{-1})} = 1.4 \times 10^{-3}$ cm.

(b)   If the copper and aluminum are removed, then $I = I_0 e^{-(610 + 1.40 \times 10^{-3})} = 0.426 I_0$. About 43% of the x-rays get through whereas 33% got through before.

14-47 $x = \dfrac{\ln 2}{\mu} = \dfrac{\ln 2}{0.18} = 3.85$ cm

This means that x-rays can probe the human body to a depth of at least 3.85 cm without severe attenuation and probably farther with reasonable attenuation.

14-49 (a)   Assume he works 5 days per week, 50 weeks per year and takes 8 x-rays per day. # x-rays = 2 000 x-rays per year and $\dfrac{5}{2\,000} = 0.002\,5$ rem per x-ray.

      (b)   5 rem/yr is 38 times the background radiation of 0.13 rem/yr.

14-51 The second worker received twice as much radiation <u>energy</u> but he received it in twice as much tissue. Radiation dose is an intensive, not extensive quantity—measured in joules <u>per kilogram</u>. If you double this energy <u>and</u> the exposed mass, the number of rads is the same in the two cases.

14-53 One rad → Deposits 10^{-2} J/kg, therefore 25 rad → 25×10^{-2} J/kg

      If $M = 75$ kg, $E = (75 \text{ kg})(25 \times 10^{-2} \text{ J/kg}) = 18.8$ J

14-55 One electron strikes the first dynode with 100 eV of energy: 10 electrons are freed from the first dynode. These are accelerated to the second dynode. By conservation of energy the number freed here, N is: $(10)(\Delta V) = (N)(10)$ or $10(200 - 100) = N(10)$ so $N = 100$. By the seventh dynode, $N = 10^6$ electrons. Up to the seventh dynode, we assume all energy is conserved (no losses). Hence we have 10^6 electrons impinging on the seventh dynode from the sixth. These are accelerated through $(700 - 600)$ V. Hence $E = (10^6)(100) = 10^8$ eV. In addition some energy is needed to cause the 10^6 electrons at the seventh dynode to move to the counter.

14-57 To conserve momentum, the two fragments must move in opposite directions with speeds v_1 and v_2 such that

$$m_1 v_1 = m_2 v_2 \qquad \text{or} \qquad v_2 = \left(\frac{m_1}{m_2}\right) v_1.$$

The kinetic energies after the break-up are then $K_1 = \frac{1}{2} m_1 v_1^2$ and

$$K_2 = \frac{1}{2} m_2 v_2^2 = \frac{1}{2} m_2 \left(\frac{m_1}{m_2}\right)^2 v_1^2 = \left(\frac{m_1}{m_2}\right) K_1.$$

The fraction of the total kinetic energy carried off by m_1 is

$$\frac{K_1}{K_1 + K_2} = \frac{K_1}{K_1 + (m_1/m_2)K_1} = \frac{m_2}{m_1 + m_2}$$

and the fraction carried off by m_2 is $1 - \dfrac{m_2}{m_1 + m_2} = \dfrac{m_1}{m_1 + m_2}$.

14-59 (a) $\Delta V = 4\pi r^2 \Delta r = 4\pi (14.0 \times 10^3 \text{ m})^2 (0.05 \text{ m}) = 1.23 \times 10^8 \text{ m}^3 \sim 10^8 \text{ m}^3$

      (b)   The force on the next layer is determined by atmospheric pressure.

$$W = P\Delta V = (1.013 \times 10^5 \text{ N/m}^2)(1.23 \times 10^8 \text{ m}^3) = 1.25 \times 10^{13} \text{ J} \sim 10^{13} \text{ J}$$

(c) 1.25×10^{13} J $= \dfrac{1}{10}$ (yield), so yield $= 1.25 \times 10^{14}$ J $\sim 10^{14}$ J

(d) $\dfrac{1.25 \times 10^{14} \text{ J}}{4.2 \times 10^{9} \text{ J/ton TNT}} = 2.97 \times 10^{4}$ ton TNT $\sim 10^{4}$ ton TNT or ~ 10 kiloton

15

Particle Physics

15-1   The time for a particle traveling with the speed of light to travel a distance of 3×10^{-15} m is

$$\Delta t = \frac{d}{v} = \frac{3 \times 10^{-15} \text{ m}}{3 \times 10^8 \text{ m/s}} = 10^{-23} \text{ s}.$$

15-3   The minimum energy is released, and hence the minimum frequency photons are produced. The proton and antiproton are at rest when they annihilate. That is, $E = E_0$ and $K = 0$. To conserve momentum, each photon must carry away one-half the energy. Thus,

$E_{min} = hf_{min} = \dfrac{2E_0}{2} = E_0 = 938.3$ MeV. Thus,

$$f_{min} = \frac{(938.3 \text{ MeV})(1.6 \times 10^{-19} \text{ J/MeV})}{6.63 \times 10^{-34} \text{ J s}} = 2.26 \times 10^{23} \text{ Hz}$$

and

$$\lambda_{max} = \frac{c}{f_{min}} = \frac{3 \times 10^8 \text{ m/s}}{2.26 \times 10^{23} \text{ Hz}} = 1.32 \times 10^{-15} \text{ m}$$

15-5   The rest energy of the Z^0 boson is $E_0 = 96$ GeV. The maximum time a virtual Z^0 boson can exist is found from $\Delta E \Delta t = \hbar$.

$$\Delta t = \frac{\hbar}{\Delta E} = \frac{1.055 \times 10^{-34} \text{ J s}}{(96 \text{ GeV})(1.6 \times 10^{-10} \text{ J/GeV})} = 6.87 \times 10^{-27} \text{ s}.$$

The maximum distance it can travel in this time is

$$d = c(\Delta t) = (3 \times 10^8 \text{ m/s})(6.87 \times 10^{-27} \text{ s}) = 2.06 \times 10^{-18} \text{ m}.$$

The distance d is an approximate value for the range of the weak interaction.

15-7   Use Table 15.2 to find properties that can be conserved in the given reactions

		Reaction 1	Reaction 2
(a)	Charge:	$\pi^- + p \rightarrow K^- + \Sigma^+$ $(-)+(+)\rightarrow(-)+(+)$ $0\rightarrow 0\ \surd$	$\pi^- + p \rightarrow \pi^- + \Sigma^+$ $(-)+(+)\rightarrow(-)+(+)$ $0\rightarrow 0\ \surd$
(b)	Baryon number:	$(0)+(+1)\rightarrow(0)+(+1)$ $+1\rightarrow +1\ \surd$	$(0)+(+1)\rightarrow(0)+(+1)$ $+1\rightarrow +1\ \surd$
(c)	Strangeness:	$(0)+(0)\rightarrow(+1)+(-1)$ $0\rightarrow 0\ \surd$	$(0)+(0)\rightarrow(0)+(+1)$ $0\rightarrow(-1)\ X$

Thus, the second reaction is not allowed since it does not conserve strangeness.

15-9   (a) $p \rightarrow \pi^- + \pi^0$ (Baryon number is violated: $1\rightarrow 0+0$)

(b) $p+p \rightarrow p+p+\pi^0$ (This reaction can occur)

(c) $p+p \rightarrow p+\pi^+$ (Baryon number is violated: $1+1\rightarrow 1+0$)

(d) $\pi^+ \rightarrow \mu^+ + v_\mu$ (This reaction can occur)

(e) $n \rightarrow p+e^- + \bar{v}_e$ (This reaction can occur)

(f) $\pi^+ \rightarrow \mu^+ + n$ (Violates baryon number: $0\rightarrow 0+1$, and violates muon-lepton number: $0\rightarrow -1+0$.

15-11   (a) $\mu^- \rightarrow e+\gamma$ $L_e: 0\rightarrow 1+0$ and $L_\mu: 1\rightarrow 0+0$

(b) $n \rightarrow p+e^- + v_e$ $L_e: 0\rightarrow 0+1+1$

(c) $\Lambda^0 \rightarrow p+\pi^0$ Strangeness $-1\rightarrow 0+0$ and charge $0\rightarrow +1+0$

(d) $p \rightarrow e^+ + \pi^0$ Baryon number $+1\rightarrow 0+0$ and lepton number $0\rightarrow 1+0$

(e) $\Xi^0 \rightarrow n+\pi^0$ Strangeness $-2\rightarrow 0+0$

15-13   (a)   In Equation 15.16, $K_{th} = \dfrac{(m_3+m_4+m_5+m_6)^2 c^2 - (m_1+m_2)^2 c^2}{2m_2}$ where m_1 is the mass

of the incident particle, m_2 is the mass of the stationary target particle, and m_3, m_4, m_5, and m_6 are the product particle masses. For $\bar{p}$ production,

$$K_{th} = \frac{(4m_p)^2 c^2 - (2m_p)^2 c^2}{2m_p} = 6m_p c^2 = (6)(938.3\ \text{MeV}) = 5\,630\ \text{MeV or } 5.63\ \text{GeV}.$$

(b)   Using Equation 15.16 for the reaction $p + p + n + \bar{n}$,

$$K_{th} = \frac{\left(2m_p + 2m_n\right)^2 c^2 - \left(2m_p\right)^2 c^2}{2m_p}$$

$$= \frac{(4)\left[(938.8 + 939.6)^2 \text{MeV}^2 c^2 - \left(4(938.3)^2 \text{MeV}^2 c^2\right)\right]}{(2)(938.3 \text{ MeV})} = 5.64 \text{ GeV}$$

15-15   Let E = efficiency in %

For Example 15.5, $E = \left(\dfrac{m_{\pi^0} c^2}{K_{th}}\right) \times 100 = \left[\dfrac{135 \text{ MeV}}{280 \text{ MeV}}\right] \times 100 = 48\%$

For Exercise 3, $E = \left(\dfrac{m_{\pi^+} c^2}{K_{th}}\right) \times 100 = \left[\dfrac{139.6 \text{ MeV}}{292 \text{ MeV}}\right] \times 100 = 48\%$

$E = 2\left(\dfrac{m_{\pi^+} c^2}{K_{th}}\right) \times 100 = 2\left[\dfrac{139.6 \text{ MeV}}{600 \text{ MeV}}\right] \times 100 = 46\%$

For Problem 13, $E = 2\left(\dfrac{m_p c^2}{K_{th}}\right) \times 100 = 2\left[\dfrac{938.3 \text{ MeV}}{5.63 \text{ GeV}}\right] \times 100 = 33\%$

15-17 $\Sigma^0 + p \rightarrow \Sigma^+ + \gamma + X$

$dds + uud \rightarrow uds + 0 + ?$

The left side has a net 3d, 2u, and 1s. The right hand side has 1d, 1u, and 1s leaving 2d and 1u missing. The unknown particle is a neutron, udd. Baryon and strangeness numbers are conserved.

15-19   Quark composition of proton = uud, and of neutron = udd. Thus, if we neglect binding energies, we may write:

$$m_p = 2m_u + m_d \tag{1}$$
$$\text{and } m_n = m_u + 2m_d \tag{2}$$

Solving simultaneously, we find:

$$m_u = \frac{1}{3}\left(2m_p - m_n\right) = \frac{1}{3}\left[2(938.3 \text{ MeV}/c^2) - 939.6 \text{ MeV}/c^2\right] = 312.3 \text{ MeV}/c^2,$$

and from either Equations (1) or (2), $m_d = 313.6 \text{ MeV}/c^2$. These should be compared to the experimental masses $m_u \cong 5 \text{ MeV}/c^2$ and $m_d \simeq 10 \text{ MeV}/c^2$.

15-21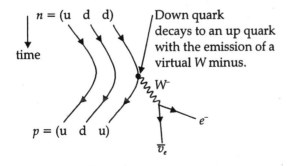

$n = (u \quad d \quad d)$

time

Down quark decays to an up quark with the emission of a virtual W minus.

$p = (u \quad d \quad u)$

W^-

e^-

$\bar{\nu}_e$

15-23   A photon travels the distance from the Large Magellanic Cloud to us in 170 000 years. The hypothetical massive neutrino travels the same distance in 170 000 years plus 10 seconds:

$$c(170\,000 \text{ yr}) = v(170\,000 \text{ yr} + 10 \text{ s})$$

$$\frac{v}{c} = \frac{170\,000 \text{ yr}}{170\,000 \text{ yr} + 10 \text{ s}} = \frac{1}{1 + \left\{ 10 \text{ s}/\left[\left(1.7 \times 10^5 \text{ yr}\right)\left(3.156 \times 10^7 \text{ s/yr}\right)\right]\right\}} = \frac{1}{1 + 1.86 \times 10^{-12}}$$

For the neutrino we want to evaluate mc^2 in $E = \gamma mc^2$:

$$mc^2 = \frac{E}{\gamma} = E\sqrt{1 - \frac{v^2}{c^2}} = 10 \text{ MeV}\sqrt{1 - \frac{1}{\left(1 + 1.86 \times 10^{-12}\right)^2}} = 10 \text{ MeV}\sqrt{\frac{\left(1 + 1.86 \times 10^{-12}\right)^2 - 1}{\left(1 + 1.86 \times 10^{-12}\right)^2}}$$

$$mc^2 \approx 10 \text{ MeV}\sqrt{\frac{2\left(1.86 \times 10^{-12}\right)}{1}} = 10 \text{ MeV}\left(1.93 \times 10^{-6}\right) = 19 \text{ eV}$$

Then the upper limit on the mass is

$$m = \boxed{\frac{19 \text{ eV}}{c^2}}$$

$$m = \frac{19 \text{ eV}}{c^2}\left(\frac{\text{u}}{931.5 \times 10^6 \text{ eV}/c^2}\right) = 2.1 \times 10^{-8} \text{ u}.$$

15-25 $m_\Lambda c^2 = 1\,115.6 \text{ MeV}$ $\Lambda^0 \to p + \pi^-$

$m_p c^2 = 938.3 \text{ MeV}$ (See Table 15.2 for masses)

$m_\pi c^2 = 139.6 \text{ MeV}$

The difference between starting mass-energy and final mass-energy is the kinetic energy of the products.

$$K_p + K_\pi = 37.7 \text{ MeV} \text{ and } p_p = -p_\pi$$

Applying conservation of relativistic energy,

$$\left[(938.3 \text{ MeV})^2 + p^2 c^2\right]^{1/2} - 938.3 \text{ MeV} + \left[(139.6 \text{ MeV})^2 + p^2 c^2\right]^{1/2} - 139.6 \text{ MeV} = 37.7 \text{ MeV}.$$

Solving the algebra yields $p_p c = -p_\pi c = 100.4 \text{ MeV}$. Then

$$K_p = \left[\left(m_p c^2\right)^2 + (100.4 \text{ MeV})^2\right]^{1/2} - m_p c^2 = 5.4 \text{ MeV}$$

$$K_\pi = \left[(139.6 \text{ MeV})^2 + (100.4 \text{ MeV})^2\right]^{1/2} - 139.6 \text{ MeV} = 32.3 \text{ MeV}$$

15-27   Time-dilated lifetime.

$$T = \gamma T_0 = \frac{0.9 \times 10^{-10} \text{ s}}{\left(1 - v^2/c^2\right)^{1/2}} = \frac{0.9 \times 10^{-10} \text{ s}}{\left(1 - (0.96)^2\right)^{1/2}} = 3.214 \times 10^{-10} \text{ s}$$

distance $= (0.96)\left(3 \times 10^8 \text{ m/s}\right)\left(3.214 \times 10^{-10} \text{ s}\right) = 9.3 \text{ cm}$

15-29 $p + p \rightarrow p + \pi^{+} + X$

$Q = M_p + M_p - M_p - M_{\pi^{+}} - M_X$

(From conservation of momentum, particle X has zero momentum and thus zero kinetic energy.)

$Q = (2)(70.4 \text{ MeV}) = 938.3 \text{ MeV} + 938.3 \text{ MeV} - 938.3 \text{ MeV} - 139.5 \text{ MeV} - M_X$

$M_X = 939.6 \text{ MeV}$

X must be a neutral baryon of rest mass 939.6 MeV/c^2. Thus X is a neutron.

15-31   (a)   The mediator of this weak interaction is a Z^0 boson.
         (b)   The mediator of a strong (quark-quark) interaction is a gluon.

15-33   (a) $\Delta E = (m_n - m_p - m_e)c^2$. From Appendix B,
                  $\Delta E = (1.008\ 665 \text{ u} - 1.078\ 25 \text{ u})931.5 \text{ MeV/u} = 0.782 \text{ MeV}$

         (b)   Assuming the neutron at rest, momentum is conserved, $p_p = p_e$ relativistic energy is

                  conserved, $\left[\left(m_p c^2\right)^2 + \left(p^2 c^2\right)\right]^{1/2} + \left[\left(m_e c^2\right)^2 + \left(p_e^2 c^2\right)\right]^{1/2} = m_n c^2$. Since $p_p = p_e$.

                  $\left[(938.3 \text{ MeV})^2 + (pc)^2\right]^{1/2} + \left[(0.511 \text{ MeV})^2 + (pc)^2\right]^{1/2} = 939.36 \text{ MeV}$

         Solving the algebra $pc = 1.19 \text{ MeV}$. If $p_e c = \gamma m_e v_e c = 1.19 \text{ MeV}$, then,

         $\dfrac{\gamma v_e}{c} = \dfrac{1.19 \text{ MeV}}{0.511 \text{ MeV}} = \dfrac{x}{\left(1 - x^2\right)^{1/2}} = 2.329$ where $x = \dfrac{v_e}{c}$

                  $x^2 = \left(1 - x^2\right)5.423$

                  $x = \dfrac{v_e}{c} = 0.919$

                  $v_e = 0.919c = 276 \times 10^6 \text{ m/s}$

         Then $m_p v_p = \gamma m_e v_e = \dfrac{(1.19 \text{ MeV})\left(1.6 \times 10^{-13} \text{ J/MeV}\right)}{3 \times 10^8 \text{ m/s}}$

                  $v_p = \dfrac{\gamma m_e v_e}{m_p c} = \dfrac{(1.19 \text{ MeV})\left(1.6 \times 10^{-13} \text{ J/MeV}\right)}{\left(1.67 \times 10^{-27} \text{ kg}\right)\left(3 \times 10^8 \text{ m/s}\right)} = 3.80 \times 10^5 \text{ m/s}$

                  $v_p = 380 \text{ km/s} = 0.001\ 266c$

         (c)   The electron is relativistic, the proton is not.

15-35   (a) $p_{\Sigma^{+}} = eBr_{\Sigma^{+}} = \dfrac{\left(1.602\ 177 \times 10^{-19} \text{ C}\right)(1.15 \text{ T})(1.99 \text{ m})}{5.344\ 288 \times 10^{-22} (\text{kg} \cdot \text{m/s})/(\text{MeV}/c)} = \dfrac{686 \text{ MeV}}{c}$

                  $p_{\pi^{+}} = eBr_{\pi^{+}} = \dfrac{\left(1.602\ 177 \times 10^{-19} \text{ C}\right)(1.15 \text{ T})(0.580 \text{ m})}{5.344\ 288 \times 10^{-22} (\text{kg} \cdot \text{m/s})/(\text{MeV}/c)} = \dfrac{200 \text{ MeV}}{c}$

(b)   Let φ be the angle made by the neutron's path with the path of the Σ^+ at the moment of decay. By conservation of momentum:

$$p_n \cos\varphi + (199.961\,581 \text{ MeV}/c)\cos 64.5° = 686.075\,081 \text{ MeV}/c$$

$$\therefore p_n \cos\varphi = 599.989\,401 \text{ MeV}/c \tag{1}$$

$$p_n \sin\varphi = (199.961\,581 \text{ MeV}/c)\sin 64.5° = 180.482\,380 \text{ MeV}/c \tag{2}$$

From (1) and (2):

$$p_n = \sqrt{(599.989\,401 \text{ MeV}/c)^2 + (180.482\,380 \text{ MeV}/c)^2} = 627 \text{ MeV}/c.$$

(c) $E_{\pi^+} = \sqrt{(p_{\pi^+}c)^2 + (m_{\pi^+}c^2)^2} = \sqrt{(199.961\,581 \text{ MeV})^2 + (139.6 \text{ MeV})^2} = 244 \text{ MeV}$

$E_n = \sqrt{(p_n c)^2 + (m_n c^2)^2} = \sqrt{(626.547\,022 \text{ MeV})^2 + (939.6 \text{ MeV})^2} = 1\,130 \text{ MeV}$

$E_{\Sigma^+} = E_{\pi^+} + E_n = 243.870\,445 \text{ MeV} + 1\,129.340\,219 \text{ MeV} = 1\,370 \text{ MeV}$

(d) $m_{\Sigma^+}c^2 = \sqrt{E_{\Sigma^+}^2 - (p_{\Sigma^+}c)^2} = \sqrt{(1\,373.210\,664 \text{ MeV})^2 - (686.075\,081 \text{ MeV})^2} - 1\,190 \text{ MeV}$

$\therefore m_{\Sigma^+} = 1\,190 \text{ MeV}/c^2$

$E_{\Sigma^+} = \gamma m_{\Sigma^+}c^2$, where $\gamma = \left(1 - \dfrac{v^2}{c^2}\right)^{-1/2} = \dfrac{1\,373.210\,664 \text{ MeV}}{1\,189.541\,303 \text{ MeV}} = 1.154\,4$. Solving for v,

$v = 0.500c$.

15-37   (a)   If $2N$ particles are annihilated, the energy released is $2Nmc^2$. The resulting photon momentum is $p = \dfrac{E}{c} = \dfrac{2Nmc^2}{c} = 2Nmc$. Since the momentum of the system is conserved, the rocket will have momentum $2Nmc$ directed opposite the photon momentum. $p = 2Nmc$.

(b)   Consider a particle that is annihilated and gives up its rest energy mc^2 to another particle that also has initial rest energy mc^2 (but no momentum initially).

$$E^2 = p^2 c^2 + (mc^2)^2$$

Thus $(2mc^2)^2 = p^2 c^2 + (mc^2)^2$. Where p is the momentum the second particle acquires as a result of the annihilation of the first particle. Thus $4(mc^2)^2 = p^2 c^2 + (mc^2)^2$, $p^2 = 3(mc^2)^2$. So $p = \sqrt{3}mc$. This process is repeated N times (annihilate $\dfrac{N}{2}$ protons and $\dfrac{N}{2}$ antiprotons). Thus the total momentum acquired by the ejected particles is $\sqrt{3}Nmc$, and this momentum is imparted to the rocket.

$$p = \sqrt{3}Nmc$$

(c)   Method (a) produces greater speed since $2Nmc > \sqrt{3}Nmc$.